Meat Me Halfway

How Changing the Way We Eat Can Improve Our Lives and Save Our Planet

BRIAN KATEMAN

 Prometheus Books

Guilford, Connecticut

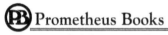
Prometheus Books

An imprint of Globe Pequot, the trade division of
The Rowman & Littlefield Publishing Group, Inc.
4501 Forbes Blvd., Ste. 200
Lanham, MD 20706
www.rowman.com

Distributed by NATIONAL BOOK NETWORK

British Library Cataloguing in Publication Information Available

Library of Congress Cataloging-in-Publication Data
Names: Kateman, Brian, author.
Title: Meat me halfway : how changing the way we eat can improve our lives
  and save our planet / Brian Kateman ; foreword by Bill McKibben.
Description: Lanham, MD : Rowman & Littlefield, [2022] | Summary: "In Meat
  Me Halfway, author and founder of the reducetarian movement Brian
  Kateman puts forth a realistic and balanced goal: mindfully reduce your
  meat consumption. The question is not how to ween society off meat, but
  how to make meat more healthy, more humane, and more sustainable"—
  Provided by publisher.
Identifiers: LCCN 2021042411 (print) | LCCN 2021042412 (ebook) | ISBN
  9781633887916 (cloth) | ISBN 9781633887923 (epub)
Subjects: LCSH: Meat—Nutrition. | Meat—Health aspects. |
  Meat—Environmental aspects.
Classification: LCC TX556.M4 K38 2022 (print) | LCC TX556.M4 (ebook) |
  DDC 641.3/6—dc23/eng/20211027
LC record available at https://lccn.loc.gov/2021042411
LC ebook record available at https://lccn.loc.gov/2021042412

For all those working tirelessly to create a more sustainable, healthy, and compassionate world

Contents

Foreword

BILL McKIBBEN

"Halfway" sounds like a compromise—but in fact this is quite a radical book and idea. Not to mention smart and elegant.

Here's the problem: We eat way too much meat. It doesn't matter whether you care mostly about the destruction of our climate, or the ravaging of rural areas by "concentrated animal feeding operations" that produce more sewage than big cities, or the cruelty to animals who must live out their lives in such conditions, or the effects on our arteries—in each of these cases we'd be far better off if our consumption fell.

For some people that means consumption falling to zero: vegetarianism or veganism are wise choices. But diets are deeply rooted in culture and in history, and for many people going cold turkey (perhaps not the best choice of words) is very hard. I've spent much of my life as an organizer traveling to remote and poor regions of the world, and I know how ingrained the appeal of animal protein can be. But each of us can easily *reduce* the amount of meat we eat.

One answer lies in meeting people where they are—not preaching to the choir but providing people with tools and resources that will help them simply cut back on the amount of animal products they consume (without it being an all-or-nothing premise), all while supporting delicious, affordable, and convenient alternatives in the marketplace. That, anyway, is the thesis of *Meat Me Halfway*. At its core, it's a message of positivity and hope, that if we all band together and do our part, we can create a sustainable, healthy, and compassionate world for generations to come.

I couldn't be more delighted to be a very small part of this movement and thank you for your support in advancing our shared mission. Our future depends on it.

Introduction

This place is the reason I turned out weird. —*Pete Davidson*

When out-of-towners think of New York City, they imagine gilded sky-scrapers and a grid of crowded streets. They don't think about Staten Island, where I was born. While Staten Island can get a bit crowded at times—with nearly five hundred thousand residents crammed into less than sixty square miles—it feels more like a distant suburb than an official borough of the largest city in the United States. There are actually over sixty neighborhoods that make up Staten Island, and locals will usually refer to a subsection of it when they mention where they're from. (I'm from Bulls Head.)

Staten Island isn't known as the hippest or most progressive of places, and it doesn't help that you need to hop on a boat to get there. Just ask Amy Schumer's character from the movie *Trainwreck*, who took the ride of shame on the Staten Island Ferry after a night of questionable decisions. The borough is perhaps most commonly known as the birthplace of Mike "The Situation" Sorrentino from the reality TV show *Jersey Shore*, for requesting to secede from New York City back in 1993 (dubbed Stexit), as the source of noxious odors from Fresh Kills Landfill (what used to be the world's largest dump), and for being the only borough in New York to vote overwhelmingly for Donald Trump . . . twice. Even the accent is slightly more pronounced than what you find in the rest of New York City. Words like "mozzarella," "calamari," and "prosciutto" (over 35 percent of Staten Island residents are of Italian descent, higher than any other county in the country) are pronounced *mooz-uh-rell, ga-la-mad,* and *pro-shoot-toe.* All this is to say that if you're from Staten Island, other New Yorkers can usually detect it. On my first day at the High School of Telecommunication Arts and Technology in Bay Ridge, Brooklyn, several students wouldn't let me say more than a few sentences before interjecting, "Are you, like, from Staten Island?" Even though each inquiry lacked any hint of condescension, I still disliked being poked at as though I belonged not in homeroom but in a cabinet of curiosities.

Mind you, the ease with which people detected my roots didn't stop me from trying to obfuscate them, particularly as I got older. When I first started online dating, I listed my location on my OkCupid profile as "New York, NY," despite living on Staten Island at the time. When rounds of messages led to in-person dates with women who lived in Manhattan or Brooklyn, I waited as long as possible before revealing what felt like a dirty secret, lest its reputation ruin my chances with them. (While I no longer feel the need to hide my Staten Island roots in shame, I can't honestly say I regret this strategy, since it just so happens I met my wife this way.) I even tried to jettison my lingering Staten Island accent, paying particular attention to words that triggered native pronunciations. During the second week of my summer internship at Columbia University, a visitor asked me where the bathroom was, and with my boss in earshot, I reflexively replied "fawth flaw." Mortified, I almost never dropped an "R" again. (I say "almost never" because it returns with a vengeance whenever I'm angry or surrounded by family. As the singer Kesha said, "We R Who We R.")

Despite its less-than-ideal reputation and occasionally embarrassing idiosyncrasies, Staten Island was a fine place to grow up. I had a fairly typical childhood as part of an American middle-class family. I lived on a placid suburban street. Kids played kickball in the neighborhood park or in the street outside. At night—aside from the occasional cacophony of crickets and car alarms—it was dead quiet. I could even make out the stars, even with the dim streetlamps switched on. On the day after Thanksgiving, our neighbors competed for the unspoken title of "Most Decked-Out Home." (Given that my family was Jewish, it was understood we weren't in the running, even if a plastic electric Hanukkah menorah in the window is a beautiful sight to behold in its own right.)

My mom, Linda, was a devoted homemaker, and my dad, Russell, was an optometrist. For fun, I would occasionally call his office and ask the receptionist if I could speak to "Dr. Kateman," finding hilarity in the fact that such a jokester at home could somehow be taken seriously at work. This was the same man who told me that during graduate school he and his friends played catch with an eyeball that once belonged to a monkey. Like so many little boys, I grew up idolizing my dad. My older sister, Jennifer, did too. He was our hero: the smartest and funniest person we'd ever met. He worked long hours to pay the bills—always until late at night and more often than not on the weekends. My happiest memories are of him walking through the front door each night, meaning we could finally sit down and eat together as a family.

And what did we eat? Meat—a lot of it. "I made meatloaf, spaghetti and meatballs, pot roast, fried fish, sometimes lamb chops or beef stew," my mom recollects. "Oh, sometimes you would help me make your favorite recipe—chicken cutlet Parmesan," she added with pride. My dad almost never prepared dinner, but when he did, it was always barbecued skirt steak on the outside grill.

Meat and dairy were also prominently featured in our breakfasts and lunches. For breakfast, particularly on the weekends, it was bacon, eggs, grilled cheese sandwiches, steak and eggs, sausage patties, and bagels and lox. For lunch, Mom often microwaved pepperoni pizza, fish sticks, mac and cheese with hamburger meat, or fried chicken nuggets. Whenever we were eating—no matter the time of day—we were eating meat. As I grew older, I continued eating this Standard American Diet (SAD)—that is, the typical diet of the majority of Americans. Though it lacks a precise definition, it's generally high in red meat, poultry, seafood, and refined carbohydrates like white bread, sugar, flour, and white rice. Similar to my friends and family, I rarely ate fruit, vegetables, whole grains, or legumes (e.g., beans and lentils). I had never even heard of quinoa.

This all changed after a single airplane flight. While traveling from New York to Montana to present undergraduate research on the environmental impacts of climate change, a classmate handed me a book called *The Ethics of What We Eat: Why Our Food Choices Matter* by Peter Singer and Jim

Manson. It was an awkward moment because I was eating a cheeseburger at the time—but I read it cover to cover. Over those few hours, I learned in vivid detail how the industrialization of animal agriculture led to cheap foods that are produced at the expense of animal welfare. Over the next several weeks I went deeper. I picked up *The CAFO Reader: The Tragedy of Industrial Animal Factories* by Daniel Imhoff, which depicted pigs confined to gestation crates, metal enclosures so small they are unable to turn around; cows enduring painful mutilations, including dehorning, castration, and branding; and drugged-up chickens selectively bred to grow so large so quickly that many of them cannot walk by the time they are slaughtered some forty days after birth. I had heard about factory farms— also known as "concentrated animal feeding operations" (CAFOs)—from which industrial meat* is born, but I had thought it was the exception, not the rule. Apparently, I wasn't alone: A 2017 survey found 58 percent of US adults think "most farmed animals are treated well," and 75 percent of US adults say they usually buy animal products "from animals that are treated humanely."[1]

I was shocked to learn that of the nine billion land animals (the number of sea animals killed is so high it's too difficult to estimate) raised for food in the United States annually, 99 percent of them are subjected to these cruel conditions. Worldwide, there are 70 billion land animals raised for food, over 90 percent of which are living in factory farms, according to some estimates. Even the pejorative term "factory farm" is a misnomer, really, because CAFOs don't share any meaningful resemblance with an archetypical farm, instead operating under an industrial factory framework, more similar to mass producing mugs or chairs. I then poured over convincing data that breeding animals for food accelerates climate change and biodiversity loss; moreover, it also increases incidences of heart disease, diabetes, obesity, and numerous cancers.

It became clear to me that eating less meat—especially industrial meat—was critical for both human health and the planet's health, so it wasn't long before I decided to become a vegetarian. I didn't personally know any vegetarians, but I thought, *How hard could it be?*

---

* By "meat," I just mean the flesh of an animal, broadly. I'll use "factory farmed" or "industrial" to refer to meat that came from a CAFO, and terms like "better meat," "higher-welfare meat," or "sustainable meat" to describe meat that didn't come from a CAFO.

I quickly realized that like most people, my food choices had been based primarily on price, convenience, and taste—not necessarily on what was good for my body or the planet. With inexpensive and delicious meaty options more readily available to me than the occasional frozen veggie burger you might find on Staten Island, I found myself "falling off the tofu train" from time to time.

Holidays and complex social situations were particularly challenging. I remember one Thanksgiving when, under pressure to honor our family by sharing turkey together, I dropped a morsel of flesh onto my plate of veggies. True to form, my big sister teased, "I thought you were a vegetarian, Brian?" A few months later, while out to breakfast at IHOP, I instinctively snagged a piece of bacon from my friend's plate as the waiter was taking it away. As a self-identifying *Jewish* vegetarian, I felt even worse.

Battling what could only be described as an identity crisis, I swore off the vegetarian label in search of one that accurately described my dietary choices. Along the way I discovered terms like *cheating vegetarian* and *lazy vegan*—both negative and self-defeating, as they focused on shortcomings rather than successes. More neutral terms like *semi-vegetarian, mostly vegetarian,* and *flexitarian* (a person who primarily eats plant-based foods but occasionally includes meat in their diet) captured my momentary eating habits—but they were also static and exclusive to people who already followed a plant-based diet. I wasn't sold on any of them.

Then, on a hot summer afternoon in Manhattan in 2014, I met my friend Tyler Alterman for lunch. I explained my conundrum to him, and in turn he shared that he had also been cutting down on meat and was having similar difficulties explaining his choices to others. Suddenly, we both realized there was a need for a positive and inclusive term for people like us, who are committed to reducing our intake of meat while inspiring others to do the same. After many brainstorming sessions, we finally came up with the term *reducetarian.*

Ever since, we have been on a mission to reduce societal consumption of meat—beginning in my hometown of Staten Island and now all around the world. I founded a nonprofit organization called The Reducetarian Foundation. I've published two other books: *The Reducetarian Solution* and *The Reducetarian Cookbook.* I've organized Reducetarian Summits in New York City, Los Angeles, Washington, D.C., and San Francisco. All

brought hundreds of leaders together to explore the most promising strategies for pushing the needle on meat consumption. I've given countless
talks and published dozens of op-eds calling for an end to industrial animal agriculture. I've secured international press for the movement, with
"reducetarian" having been translated into over twenty languages, from
Spanish (*reducetariana*) and Italian (*reducetariani*) to German (*reduce-
tarianer*) and Turkish (*redüktanizm*). I've even produced a documentary
on which this book is largely based.

All this activism begs the obvious question: Have the efforts paid off?

On one hand we are witnessing a shift in how people think about meat.
In 2016, Oprah Winfrey acknowledged Meatless Mondays as something
she "can do" and urged her millions of followers on Twitter to join her. In
2017, reality television personality Kylie Jenner announced via Snapchat
that she had adopted a vegan diet, joining celebrities like Ariana Grande,
Woody Harrelson, Ellen DeGeneres, and politicians Cory Booker, Bill
Clinton, and Al Gore in touting the benefits of swapping out meat for
plants. In 2019, which the *Economist* deemed "The Year of the Vegan,"
Beyoncé encouraged her fans to incorporate full or partial veganism into
their daily routines for a chance to win tickets to her and Jay-Z's shows.
Though polls vary widely (and should be treated cautiously), they suggest
a fast-growing segment of the general public is interested in reducing
their meat consumption. In 2018, more than one in three Brits said in a
consumer survey that they had recently reduced their meat consumption,
up from 28 percent in 2017. Similar trends have been reported in US consumer research, as 36 percent said in 2021 that they are eating less meat
than they did a few years ago. Indeed, the rise in popularity of plant-based
eating is an extremely encouraging trend, one I am cautiously optimistic
will continue for the foreseeable future.

Nevertheless, the vast majority of people still eat meat regularly. Until
recently I lived on the Upper East Side of Manhattan—one of the wealthi-
est and most privileged neighborhoods in the world—and for every
sweetgreen, Just Salad, Chop't, and Dig Inn (which all have plant-forward
menus), there are countless more meat-centric eateries. Just a few blocks
north in East Harlem, this reality is even more apparent, with a Taco Bell,
KFC, or Wendy's seemingly on every corner. (Researchers often refer to
these areas, where fast food and junk food outlets outnumber healthy

ones, as "food swamps.") I saw similar landscapes in Los Angeles, in Providence, Rhode Island, and in New Jersey, where I live today. Despite my parents' health issues (my mom has struggled with her weight and my dad has type 2 diabetes) and their familiarity with the issues surrounding factory farming, they continue to consume some 225 pounds of meat annually, or roughly what the average American now eats. To put this number in perspective, my parents' parents likely ate what the average American ate in 1950—approximately 125 pounds of meat annually.

This isn't to suggest I haven't had successes farther from home; on the contrary, I've read thousands of emails, comments, and posts with inspiring stories of people successfully cutting back on meat. Each of these anecdotes brings me tremendous joy and reminds me that my work is worth it, despite how difficult it is. Nevertheless, as the saying goes, the proof is in the pudding, and right now, most of the puddings in the world aren't vegan.

Rather than regurgitate what many people already know about the impacts of factory farming or otherwise build a case for veganism, I started writing this book with three examinations.

The first: Why do we love meat so much? When and why did we start eating it? Why do we eat so much more now compared to thousands of years ago? Or even dozens of years ago? I know that evolution and the growth of factory farms had something to do with all this, but I wanted answers from experts about why meat consumption has skyrocketed throughout the past century. Really, I should say "clues to the answers," because like any highly complex topic, it's extremely difficult to make absolute sense of it all. Teasing out whether one event is the direct result of another is as much art as it is science.

Disclaimer aside, my hope was that a detailed picture would emerge from this exploration, one that would inform my second set of questions: Despite all we know about meat today, why do we continue to eat so much of it? Is it our culture? Our biology? The chemistry of meat? What roles do the economy and politics play? Or, perhaps, is the animal agriculture industry itself responsible because of clever advertising and generous government subsidies?

Finally, if meat truly is here to stay, how can we make it more sustainable, humane, and healthy? There are three predominant alternatives.

First, we can "go back to the land" and push for higher-welfare, more eco-friendly, and ostensibly healthier meat sourced from local, independent farmers. Other activists believe the future lies in "meat" made from plants rather than animals. Finally, perhaps the answer lies neither in farm country nor in plants, but in Silicon Valley, where scientists are developing cell-cultured meat: nonsentient animal flesh grown without creating the rest of the animal, meaning no skin, no bones, no feathers. While supporters of each proposal agree that the current factory-farm model must go, that is about where the consensus ends. Unfortunately, this community of well-meaning activists is teetering on the brink of civil war, all because we can't agree on the "ideal" meat of the future.

Ultimately, this book is about our complex relationship with meat. It's less about the Christmas ham or the pig that it was once a part of and more about the people at the table celebrating while eating it, or while abstaining from eating it. It's about how we define ourselves by the meat we eat today and how our definitions will evolve as meat of the future does too. Most of all, it's about how we might create a more compassionate, sustainable, and healthy world—a world we'd all like to see.

I

RISE OF CARNIVOROUS AMERICA

1

The Birth of the Omnivore

Hey, Barn, you like your steak rare? —*Fred Flintstone*

About a decade ago I attended a summer pool party in New Jersey with my now wife, Isabel, and a dozen of her coworkers and friends. Isabel and I had only recently started dating, and I was a bit nervous about meeting her friends; it had been about a year since the last time I'd eaten meat, and expecting only traditional barbecue fare, I brought a sad, soggy package of veggie burgers with me. Fortunately, thanks to Isabel, our gracious host Alice already knew I was vegetarian, so she'd purchased some much nicer veggie burgers for me in advance.

After a few hours of frisbee, swimming, and booze, it was lunchtime. Alice asked each of her guests what they wanted to eat: hamburgers, hot dogs, steaks, chicken cutlets, or veggie burgers. After collecting everyone's order, she fired up the grill and a tantalizing aroma filled the air. It was a familiar smell, a heavenly concoction I instantly associated with smoked meat. I knew intellectually it wasn't something I wanted to crave, even internally, but I found myself practically salivating over it anyway. It was distressing to realize just how intense the craving was; I hadn't experienced cravings this powerful in a long time. We all took our seats as the host placed platters of food at the center of our table. I reluctantly fashioned a veggie burger with some fixings while everyone else eagerly assembled a plate of meat. I normally love a good veggie burger, but I knew even before I bit into this one that it wasn't going to save me from this primal tailspin. After wolfing it down, I was even more irritable and dissatisfied than before.

Finally, I couldn't take it any longer. Nonchalantly, I grabbed a juicy, delicious hamburger and then quickly devoured it. I hoped nobody would notice, but of course everyone did, and an uncomfortable conversation about my hypocrisy and our seemingly ingrained pull toward meat ensued. That wasn't the last time I ate meat, and it certainly wasn't the last time I yearned for it. Where does this deep-seated lust for meat originate? Is it cultural—perhaps associated with happy childhood memories of grilling burgers with my dad in the backyard? Or is there something more primal going on?

The answer is complicated. According to science journalist Marta Zaraska, author of *Meathooked: The History and Science of Our 2.5-Million-Year Obsession with Meat,* for tens of millions of years—long before *Homo sapiens* first began walking the earth—our distant ancestors primarily ate plants. Consider *Purgatorius,* the earliest common ancestors of primates, who hit the Mesozoic scene somewhere between fifty-five and eighty-five million years ago. More closely resembling today's squirrels or lemurs than gorillas or chimpanzees, *Purgatorius* emerged in tropical or subtropical forests after some small, shrew-like, insect-eating mammal climbed into the trees, perhaps in search of some grub. This was the start of an extraordinary transition, with the descendants of this arboreal explorer coming to rely substantially on fruits, leaves, and flowers from the canopy—a change that set the stage for the emergence of the primate order. From tiny monkeys to large apes, they developed a host of superhero-like adaptations over time—including enhanced depth perception, acuity, and color vision—to help them discern the presence of plant parts that ornamented trees, shrubs, and herbs.

As more time passed, primates diverged into various lineages, and some continued to eat these plant foods, while others added new ones, like seeds and nuts, into their diets. Even when the hominin (early human) lineage formed with the last common ancestor of chimpanzees and bonobos (our closest living evolutionary cousins) in Africa about five to eight million years ago, plant-based eating had not yet gone out of style. Take, for example, *Orrorin tugenensis*, one of the earliest hominins on record, who lived just under six million years ago. *Orrorin tugenensis* sported small canines, low rounded molars, and thick enamel, suggesting a diet of mainly leaves, fruit, seeds, roots, nuts, and insects.

It isn't until about three to four million years ago that evidence of a possible wrinkle in plant-based eating begins to show up in the fossil record. In 1994, for example, scientists discovered a remarkable four-million-year-old skeleton in Ethiopia belonging to *Ardipithecus ramidus*. Its teeth, which had neither very thick nor very thin enamel, reveal *Ardipithecus ramidus* likely followed a more generalized, omnivorous diet, one that may have included some small mammals in addition to new plant foods like grasses and sedges. Similarly, chemical analyses of the teeth of another hominin species, *Australopithecus africanus*, who roamed the earth about three million years ago, suggest the species' diet also may have included a small amount of meat. All told, early hominins may have had a diet similar to modern chimpanzees, who on average eat meat fewer than ten days per year.

Any substantial uncertainty concerning meat-eating in hominins disappears about 2.5 million years ago, slightly before the arrival of the genus *Homo*. This is when fossilized bones with cuts and percussion marks—clear signs of butchery—as well as stone tools appear in the fossil record. For example, in the 1990s, at a site in Bouri, Ethiopia, paleontologists discovered 2.5-million-year-old cranial and tooth remains from a single individual whom they dubbed *Australopithecus garhi*, after the local people's word for "surprise." Also discovered were arm and leg bones from several other individuals that couldn't reliably be assigned to a species. Buried next to one of these nebulous leg bones were catfish and antelope remains, the latter of which showed definitive cut marks from stone tools. Scattered nearby were other antelope and horse bones with similar telltale signs of butchery, including a lower jaw with curved cut marks toward the back, suggesting the tongue had been cut out, as well as leg bones fractured at both ends, indicative of marrow extraction.

In 1992, at a site in Gona, Ethiopia, paleontologists unearthed not just fist-sized hunks of rock for pounding, but also more than three thousand of the first known manufactured stone tools: sharp flakes created by striking hard stones against rocks. A more recent 2018 study reported the discovery of about 250 similarly crafted tools, dated as far back as 2.4 million years ago, from a site in Ain Boucherit, Algeria. They were found alongside nearly three hundred bones of animals, about two dozen of which had cut marks. These then technological marvels allowed our ancestors to slice apart a wide range of dead animals, from gazelles to elephants, accessing for the first time the nutrient-dense components within.

How did these early humans gain access to these carcasses? Scientists continue to debate two possible scenarios. The first is the scavenger hypothesis, which proposes that early humans simply scavenged the remains left by larger carnivores like lions and saber-toothed cats. Subsumed within this hypothesis is the question of whether early humans were solely "passive" scavengers, or if they were also "confrontational" or "power" scavengers who would have actively competed with preying carnivores and other scavengers, such as hyenas and vultures. The second major theory is the hunter hypothesis, which argues that early humans were not scavengers, but active, cooperative hunters who used intricate ambushing techniques to kill gazelles, antelopes, wildebeests, and other large animals. There is archeological evidence in favor of each overarching hypothesis, so it hasn't been possible to conclusively determine which one is correct.

While we may never know definitively if early humans were scavengers, hunters, or somewhere in between, experts agree that over time they engaged in hunting more frequently and with increasingly sophisticated practices. Beginning approximately 1.6 million years ago, early humans created new kinds of tools called hand axes—large, oval-shaped multipurpose tools laboriously shaped by striking flakes off stones. Some experts posit hand axes were used as hunting projectiles or as "killer Frisbees,"[1] bombarded at a herd of animals to stun one of them. About 400,000 to 500,000 years ago, soon after our evolutionary cousins *Homo neanderthalensis* began walking the earth, humans learned to craft throwable wooden spears, enabling them to hunt from a somewhat safer distance than was possible with earlier weapons.

This discussion raises an important question: What prompted our ancestors to begin to hunt for and to eat meat in the first place? According to Briana Pobiner, a paleoanthropologist at the Smithsonian's National Museum of Natural History, "Some ideas have to do with potential changes in climate, where foods like fruit and different plants and other things that we may have eaten maybe became less available."[2] In response to shortages in traditional fanfare, early humans simply began eating meat as a way to survive.

Although there is considerable debate over the particular environmental circumstances that sparked this dietary transition, most experts agree it played a crucial role in human evolution. First and foremost, eating meat may have helped facilitate the development of our relatively big brains. For comparison, *Australopithecus afarensis* and *Australopithecus garhi* both had an average brain size of approximately 450 cubic centimeters—about the size of modern chimpanzees and orangutans. *Homo habilis*—one of the "missing links" between apes and humans—had an average brain size of approximately 600 cubic centimeters, slightly larger than modern gorillas. *Homo erectus*, who existed some 1.8 million years ago, had an average brain size of approximately 900 cubic centimeters, while today's humans average about 1,350 cubic centimeters.

Without meat—which is chock-full of calories and fat—it's possible we wouldn't have developed such big brains. That's because our big brains use a lot of juice: Despite making up only 3 percent of our weight, they consume 20 percent of our body's energy. Early humans would need to spend an impractical amount of time on foraging and feeding on low-calorie plant foods to fuel such a large organ.

This may also explain why our guts are much smaller than other hominins. According to proponents of the so-called expensive tissue hypothesis, digesting fewer bulky plant fibers would have allowed these omnivorous humans to evolve much smaller intestinal systems, freeing up energy for a resource-intensive brain. Meat-eating may also have made our teeth smaller and sharper, since flesh requires much less chewing than most plants. For example, in one study, researchers recruited twenty-four individuals and fed them samples of three kinds of plant foods—jewel yams, carrots, and beets—as well as raw goat. Then they used electromyography sensors to measure how much energy it took for head and jaw muscles to chew and swallow each food.[3]

Here's what the researchers found: "If meat comprised one-third of the diet, the number of chewing cycles per year would have declined by nearly two million (a 13 percent reduction) and total masticatory force required would have declined by 15 percent. Furthermore, by simply slicing meat and pounding [plants], hominins would have improved their ability to chew meat into smaller particles by 41 percent, reduced the number of chews per year by another 5 percent, and decreased masticatory force requirements by an additional 12 percent."[4] On average, the study found they require upward of 46 percent less force to chew and swallow processed meat. In numerous other ways—from our large and tall bodies to the thinning of our body hair—the transition from eating primarily plant foods to regularly eating meat may define the physical nature of our very existence.

There are some researchers who find this meat-centric view of human evolution to be incomplete. Chief among them is Richard Wrangham, a primatologist who argues in his book *Catching Fire: How Cooking Made Us Human* that "meat eating accounts smoothly for the first transition, jump-starting evolution toward humans by shifting chimpanzeelike australopithecines into knife-wielding, bigger brained habilines . . . [but] does not solve a key problem concerning the anatomy of *Homo erectus*, which had small jaws and small teeth that were poorly adapted for eating the tough raw meat of game animals. These weaker mouths cannot be explained by *Homo erectus*'s becoming better at hunting. Something else must have been going on."[5] That something else, Wrangham explains, is cooking. Because of our ill-equipped anatomy, raw meat would not have provided enough calories to carry out our second major evolutionary transition: from habilines to *Homo erectus*. Instead, cooking both meat and plants increased the amount of energy our ancestral bodies were able to obtain, leading to larger brains and bodies and smaller teeth and guts. As an example, when eggs are cooked, their protein value shoots up by as much as 40 percent compared to when they are raw.

Many scientists disagree with this "cooking hypothesis," noting that archaeological evidence of controlled fire is both tenuous and sparse before five hundred thousand years ago, and nearly nonexistent before one million years ago. The unequivocal evidence of habitual fire dates to about four hundred thousand years ago—long after the major evolutionary changes in our anatomy. Instead of cooking, many experts favor the idea that increased use of other processing techniques, such as pounding or cutting food using stone tools, was likely sufficient for facilitating this second evolutionary transition. At the end of the day, the debate rages on as to whether it was raw meat or cooked food that propelled the decisive leap in human evolution, but most scientists agree that without eating *some* animal parts, cooked or otherwise, humans would look and behave far differently than we do today.

While we may never know precisely how much meat ancient humans ate, we do know that it was very different from the meat we eat today, notwithstanding what the "paleo" fad diet may fancifully claim. In fact, ancestral diets varied dramatically. Some may have included high amounts of meat, some were predominantly plant based, while others fell somewhere in between. After all, there wasn't a single ancestral population—rather, there were many different populations across space and time with unique biological constitutions who encountered different environmental circumstances.

Consider the diet of Neanderthals. A 2016 study by researchers at the University of Tuebingen analyzed the relationship between environmental data and microscopic wear patterns on tooth surfaces, finding that during the Ice Age, when the climate was constantly fluctuating, Neanderthals altered their diets in response to changes in food resource availability. During cold spells, Neanderthals subsisted almost entirely on meat. During more temperate times, they likely supplemented their diet with plants, seeds, and nuts. The data also suggest early humans displayed even greater dietary flexibility than Neanderthals, finding ways to include fruits and vegetables in their diets even when extracting them was difficult. This difference in dietary flexibility may even explain why Neanderthals went extinct and humans survived. As paleoanthropologist Chris Stringer, professor and merit researcher at the Natural History Museum in London, told me, the hallmark of being human is our ability to adapt to many different habitats by eating a variety of foods. "They took what they could and there was no single paleo diet," Chris explained. "It was a varied diet, and at all times in the past certainly it would have included for humans probably some meat but often not a lot of meat and a lot of vegetable foods."[6]

Sometime between forty thousand and one hundred twenty thousand years ago, our species began to venture out of its birthplace—most likely East Africa near modern-day Ethiopia—and migrate all over the world, drawing upon this flexible omnivorous diet and the many adaptations it fueled along the way. In their path, humans left evidence of an unprecedented ability to think and plan for the future—what theorists of early human cognition call "higher-order consciousness." Indeed, by forty thousand years ago, humans exhibited fully modern behavior, including engaging in music, self-ornamentation, trade, burial rites, and figurative art. This enhanced ability to carry language and symbolic culture would set the stage for a new invention, one that would allow humans to exert much greater control over their meat-eating ways.

2

The Creation Story

And if we reflect on the vast diversity of the plants and animals which have been cultivated, and which have varied during all ages under the most different climates and treatment, we are driven to conclude that this great variability is due to our domestic productions having been raised under conditions of life not so uniform as, and somewhat different from, those to which the parent species had been exposed under nature. —*Charles Darwin*

Several years ago, Isabel and I visited the Humane Society of New York to adopt a dog. It was Isabel's idea, not mine. She was very eager to add a new member to our little family, but as for me . . . not so much.

It wasn't that I'd never liked the idea of having a companion animal. In fact, as an unruly teenager, I pulled out all the stops to have one. When my mom told me I wasn't allowed to bring any furry friends into the house, I defied by enlisting my sister to purchase a kitten for me at a nearby pet store. We named him Simba, and my mom not only agreed to let him into the house, but she eventually grew to love him as much as I did. (I'd be remiss if I didn't take this opportunity to remind readers to "adopt, don't shop," for I hadn't known any better back then.)

Yet as we gazed at those hopeful canine faces, I didn't feel ready for the responsibility. As I reminisced about carefree days of staying out late with no one else to feed or walk, a staff member brought out a four-pound, four-month-old female terrier-mix named Tobey. The poor thing was terrified, and I was too. I never thought we'd adopt a *puppy*, who are far more demanding than adult dogs. In the minutes that followed, Tobey warmed up to Isabel and me. As we were lying on the floor, she went back and forth, licking our noses. I knew Tobey and Isabel had won and I had lost—she would soon be going to her forever home. It took a couple of weeks, but eventually unbridled joy and affection replaced my reservations about Tobey. I had fallen in love with everything about her, including the way she snuggled with us in bed, the sound of her protective bark, her penchant for chasing squirrels, and even the smell of her fur. It felt like she was put on this planet to be with us.

Casting aside allusions to cosmic intervention, that sentiment isn't too far from the truth. After all, Tobey (and the latest canine member of our family, Cooper), and domesticated animals like her, was essentially created by humans. This realization made me wonder: When and under what circumstances did we achieve dominion over animals, and how did such power transform our relationship to meat?

Historically, the dynamics between humans and animals were marked largely by opposing interests, such as competition for limited resources and avoiding predation. For millennia, other than occasional clashes during confrontational scavenging and hunting, humans and animals lived separate lives. All of this started to change between fourteen thousand and forty thousand years ago, when wolves and humans entered into a new relationship known as domestication—the process by which wild animals and plants become "tamed" or adapted in a way that makes it easier for humans to consume or cooperate with them.

Researchers have different ideas about how it first happened, nearly all within the framework of what is referred to as the "commensal pathway." First, proponents of the so-called dumpster theory argue that wolves began approaching hunter-gatherer camps in search of food, and over time the less aggressive ones befriended humans. In exchange for tossing scraps to these affable canines, humans benefited from wolves, who warned them about animal and human invaders who approached at night. An alternative view is that wolves and humans helped one another gain access to meat. After wolves used their keen senses to track, pursue, and corner prey, humans used their weapons and coordinated know-how to kill it. The wolves feasted on the flesh and bones, and then the humans scavenged the remains—a win-win. Finally, some researchers believe that young wolves separated from their pack were adopted into camps as companion animals. No matter which hypothesis (or combination of hypotheses) is right, one thing is for certain: Individual wolves with traits, such as tameness, that aided in their own survival and reproduction increased in the population. Over many generations, the domestic dog emerged.

It wasn't until humans became largely sedentary some twelve thousand years ago that the domestication of animals went into overdrive. One of the first of these semi-stationary settlements originated in Mesopotamia, an ancient region between the Tigris and Euphrates Rivers in modern-day Iraq, Kuwait, Syria, Turkey, and Iran. Mesopotamia and other settlements like it were all in the quarter-moon-shaped region in the Middle East known as the "Cradle of Civilization." The Fertile Crescent was abundant in many wild animals, including gazelle, deer, cattle, boar, horses, goats, and sheep, as well as in many edible plants including seeds, leaves, fruits, tubers, cereals, and pulses. True to its name, the Fertile Crescent allowed its human inhabitants to hunt and gather all they needed with only short trips from base camps. Over time, these camps settled into log-framed buildings partly dug into the ground, known as pit houses, where inhabitants stored surpluses of wild grains for use throughout the year. Many experts think these grain stores facilitated the domestication of cats approximately ten thousand years ago, where wildcats first settled into a mutually beneficial relationship with humans, patrolling for mice and rats, which were attracted to crops and other agricultural by-products.

Goats were likely the first animals to be domesticated for consumption, between approximately nine thousand and eleven thousand years ago, followed by sheep and then cattle.

The domestication of these animals was likely accidental, as humans experimented with methods of hunting that might grow the numbers of these prey in the wild, perhaps out of desperation to reverse plummeting numbers of animals. Experts sometimes call this the "prey pathway" to domestication. Over time, these game-management strategies developed into herd-management strategies, where humans began to control the animals' movement, feeding, and reproduction. As for pigs, some research suggests their forebears, wild boar, were domesticated in Mesopotamia approximately nine thousand years ago via a combination of mutually beneficial and prey pathways, possibly helping to dispose of human table scraps and waste products in addition to being hunted in the wild. In time, humans would domesticate goats, sheep, cattle, and pigs for additional uses, including the production of wool for clothing, hides for tent shelters, and milk.

Humans may not have necessarily planned to domesticate goats, sheep, cattle, and pigs, but once they did, the sky was the limit. Soon came the more intentional domestication of horses, donkeys, camels, and other animals involved in all sorts of uses, including meat, hides, milk, draft, traction, and transport. Domestication happened not only in the Fertile Crescent but also in South Asia and Africa. Pigs, meanwhile, were also domesticated in East Asia and Western Europe. Although the exact timing, origin, and sequence of events involving chickens are complex and uncertain, they are likely descended from red jungle fowl—brilliantly colored birds with gold, red, brown, dark maroon, orange, green, gray, and white plumage—and possibly three closely related species that bred with it, and were domesticated in Southeast Asia, China, and possibly India between four thousand and eight thousand years ago.

Civilizations that embraced domestication and agriculture had several huge advantages over their nomadic, largely plant-based counterparts. For one, domesticated animals provided ready access to dietary fat. Aside from wild nut trees, such as pistachios and almonds, few plant-based sources provided large amounts of fat calories. Animal fat allowed humans to pack on body weight to help them survive during leaner times, and animals could also be traded with other groups to maximize the sharing of food. The ability to produce a reliable surplus of both plants and meat beyond the immediate needs of daily subsistence were jointly important for allowing humans to stay in one place for long periods of time.

Second, as settlements became larger and permanent, women gave birth in more rapid succession. Fueled in large part by progress in these subsistence techniques, as well as major food-storage and food-transforming technologies that emerged with them, such as pottery, fermentation, and salt as a preservative, the human population increased from approximately four million people in 10,000 BC to five million by 5,000 BC, fourteen million by 3,000 BC, and to fifty million by 1,000 BC. This new system was marked not only by the development of the first farm communities, but also the development of densely populated cities and the suite of innovations that came with them, including the basis for centralized administrations and political structures, writing, specialization and division of labor, art and architecture, and property ownership—in short, modern civilization as we know it.

Researchers do not know the extent to which meat from domesticated animals nourished these ancient civilizations, from Mesopotamians to the Sumerians in Eridu to the Egyptians in Memphis; there just aren't enough archaeological sites filled with ancient plant and animal remains to give us a clear answer. Like with the humans who lived during the Paleolithic period, how much we ate undoubtedly varied widely depending on the place and time. These early civilizations were now composed of people divided by wealth, class, occupation, and ethnicity, and their diets varied accordingly; social status, in other words, had become a new factor in determining who ate what and how much of it. For example, ancient Egyptian priests offered the gods three square meals a day, including coveted food like beef and goose; since the gods likely never took them up on their offers, the priests ate the food with their families. Laborers, in contrast, likely only ate two meals a day, only one of which included meat. (The builders of the famous Giza pyramids were an exception: Piles of animal bones near a human cemetery suggest more than four thousand pounds of meat from cattle, sheep, and goats were slaughtered every day to fuel the builders.)

Once these animals were permanently domesticated, cultural contacts, trade, migration, and territorial conquest ensued throughout the world. Within a few thousand years, domesticated animals would arrive in the Americas alongside European colonists with healthy appetites for meat, altering the lives of those who had already been there and laying the groundwork for an existence centered on animal flesh.

3

Old Habits, New World

I'm more American than apple pie. I'm like apple pie, with a hot dog in it.
—*Stephen Colbert*

When I was a kid, my dad would take me to a baseball game once or twice a year. Although we sometimes went to see the Yankees or Mets, we usually went to the Richmond County Bank Ballpark to cheer on the Staten Island Yankees, a minor league team. Neither of us were big baseball fans; we didn't know the names of most of the players or watch games on television, nor did we particularly like big crowds and loud music. Instead, we loved being there for one simple reason: the food.

From peanuts and Cracker Jacks to soft-pretzels and nachos and cheese, we reveled in junk food glory. Hot dogs were the main culinary attraction for us, and we certainly weren't alone: According to the National Hot Dog and Sausage Council, during a typical baseball season, fans consume some twenty million hot dogs. Indeed, not unlike eating popcorn at the movie theater, part and parcel with America's Pastime is the almighty hot dog. Much like standing and removing your cap for the National Anthem prior to the first pitch, eating a hot dog between innings is just what we did. Years later, pondering this tradition made me wonder: When and how did we start to associate eating meat with being a "real" American?

The story begins not on the baseball diamond, but thousands of years earlier—long before Christopher Columbus made landfall in the Bahamas in 1492. Most researchers estimate that by at least fifteen thousand years ago, the first Americans migrated in small groups from Siberia. The most popular theory is that these intrepid humans traveled across the Bering Land Bridge, which connected Russia and Alaska when the most recent ice age lowered sea levels. Over thousands of years, they spread southward and eastward across North America and beyond into South America.

By first hunting the abundance of large game—think mammoths, mastodons, and giant sloths—and then subsisting on a broader array of plants and smaller animals, the human population in North America increased tenfold, from about one hundred thousand people to one million, between three thousand and nine thousand years ago. Using the same agricultural techniques that their ancestors honed in Mesopotamia, the natives in Central America began growing squash, beans, and corn. The cultivation of these crops expanded their food supply and fueled the growth of complex civilizations. Over the next several thousand years, numerous sophisticated cultures blossomed throughout the Americas, from Hohokam and Anasazi in the American Southwest to the natives who lived in Cahokia (across the Mississippi River from present-day St. Louis), leaving behind a trail of artifacts indicative of their rich lives, including immense complexes known as "great houses," and massive, intricately engineered mounds in Cahokia.

On the other side of the globe, during the fifteenth and seventeenth centuries, Europeans developed maritime technology that allowed them to unleash their latent ambition to explore and dominate the world. In what became known as the Age of Discovery, the Portuguese were the first to sponsor expeditions in search of spices, gold, and uninhabited land. Starting in about 1419, under the direction of Prince Henry the Navigator, Portuguese ships traveled along West Africa's coast carrying spices, gold, and enslaved people from Asia and Africa to Europe. Portugal wasn't the only greedy explorer; other European nations, especially Spain, also coveted the wealth of the so-called Far East. By the end of the fifteenth century, Spain's "Reconquista"—the expulsion of Muslims and Jews out of the kingdoms of Aragon and Castile after nearly eight hundred years of war—was complete, and the nation, now led by King Ferdinand II of

Aragon and Isabella I of Castile, turned its attention to exploration and conquest.

On August 3, 1492, Christopher Columbus and a crew of more than eighty sailors set sail from the Spanish port of Palos, and, after making a pit stop in the Canary Islands, stumbled upon an island Columbus christened San Salvador. Although Columbus thought the island was in the East Indies (hence why he mislabeled the natives "Indians"), it was actually somewhere in present-day Bahamas. Afterward, he landed on the northeast coast of Cuba, and then the northwestern coast of Hispaniola (present-day Haiti and the Dominican Republic).

While Columbus settled to find "pearls, precious stones, gold, silvers, spice, and other objects and merchandise whatsoever,"[1] he suspected sugar—native to Southeast Asia—might grow exceptionally well in the rich soil of the West Indies. The Spanish began building small but viable plantations in the Caribbean, and by the 1530s they were proving to be extremely profitable. To accelerate the growth of sugar production, the Spanish sought labor, and that meant enslaving the Taíno—native populations with ancestry in South America who had thrived in the region for over a thousand years—but virulent diseases like smallpox, measles, and influenza brought by colonists nearly wiped them out.

Columbus marveled not only at the presence of the Taíno but also the absence of domesticated animals. Unlike back home, there weren't any pigs, cattle, or sheep. Aside from dogs "that did not bark," wrote Columbus in his journal, and possibly parrots that were in such great quantity they "concealed the sun,"[2] the Taíno did not keep any other domesticated animals, and they still depended heavily on plants, both cultivated ones like sweet potatoes, squash, and peppers, and wild ones like cassava. They also hunted wild animals, mainly fish, snakes, and birds. By and large, this was true for natives throughout the Americas. All told, domesticated animals in the New World included only two birds, the turkey and the Muscovy duck; a medium-sized rodent (guinea pig); and two camelids, the llama and alpaca.

Nonetheless, Columbus was determined to eat like a European. So, on his second voyage in 1493, in addition to bringing along one thousand colonists to increase the settlement on Hispaniola and convert the natives to Christianity, he also brought domesticated animals, including pigs, cattle, sheep, and goats. These animals were let loose throughout the Caribbean to live and reproduce on their own with little oversight, and within a few years the islands were overrun.

Later Spanish conquistadores, including Hernando de Soto and Pedro Menendez de Aviles, did much of the same during voyages to the Florida peninsula. In 1539, for example, De Soto brought to Tampa Bay 600 colonists, 13 pigs, 230 horses, and many dogs. Over the next three years, he traveled with these animals and their progeny throughout the southeastern United States, where many likely escaped and multiplied in the wild. As one of the first explorers to give early American colonists a taste for pork, de Soto has been dubbed "the father of the American pork industry."[3] Pedro Menendez de Aviles, who established a Spanish colony in St. Augustine, Florida, in 1565, brought more than 100 horses, as well as sheep, hogs, and cattle. All told, each time explorers ventured into the Americas, they seeded the landscape with nonnative animals who reproduced in droves, creating dramatic long-term effects on the fauna and flora, not to mention the diets of the native people, slaves, and colonists.

Explorers brought domesticated animals to the Americas largely because they believed eating European food would keep them healthy. Explaining to King Ferdinand and Queen Isabella why so many of the European settlers in the Caribbean had fallen sick and died, Columbus blamed the "change in water and air," and claimed the deaths would cease once the colonists ate "the foods we are accustomed to in Spain."[4]

Upon arriving in the Americas, the Spanish quickly noticed the natives looked physically different in many ways. They were darker skinned and beardless, had straight hair, and, in the eyes of the Spanish, were generally timid and deceitful. To explain why the natives looked and behaved so differently from them, the Spanish drew upon the so-called humor theory, which theorized that good health required a delicate balance of four "humors"—blood, phlegm, black bile, and yellow bile—which could be disturbed by a change in climate or food. Believing they would devolve into the natives if they ate foods from the New World, the Spanish put considerable effort into bringing European domesticated animals with them as they colonized the Americas. As explained in the early seventeenth century by the Spanish Dominican Gregorio García, the constitution of any Spaniard was unlikely to be lost provided they ate "good foods and sustenance such as lamb, chicken, turkey, and good beef, bread, and wine, and other nourishing foods."

Like the Spanish, English, Dutch, French, and other European settlers were well acquainted with the concept of humor theory and assumed any successful attempts at colonization would include farming the livestock of their homeland. In particular, wealthy English aristocrats were renowned for their extraordinarily high levels of meat consumption, a practice they had no interest in abandoning upon arriving in the New World. Simply put, the higher up in society you were, the more meat you ate. At the extreme, one estimate suggests a whopping 80 percent of the diet of Henry VIII's courtiers consisted of meat. As Susanne Groom details in her book *At the King's Table: Royal Dining through the Ages*, a typical first course on one of Henry VIII's "fast days" (where red meat was eschewed) included "soup, herring, cod, lampreys, pike, salmon, whiting, haddock, plaice, bream, porpoise, seal, carp, trout, crabs, lobsters, custard, tart, fritters and fruit."[5]

The meat the very rich enjoyed was fresh all year round, as they could afford to keep animals alive during the winter and kill them when required. Meanwhile, experts disagree on exactly how much fresh meat the bourgeois ate during the sixteenth and seventeenth centuries, but available evidence suggests it was undoubtedly a luxury in most homes; instead, people largely relied on salted and smoked meats that could be preserved. Still, as scholar Jeffrey L. Forgeng writes in *Daily Life in Stuart England*, "Even for those of moderate means, poultry (especially ducks, geese, turkeys, and pigeons), eggs, dairy products, pork, mutton, fish, and shellfish were common.... Foreigners traveling to England often remarked on the ubiquity of meat in the English diet, even for those of limited incomes."[6] The destitute, who constituted a huge swath of English society during this time, likely ate very little meat, instead subsisting on low-quality bread and pottage—a thick soup or stew made by boiling vegetables, grains, and what few meaty scraps they could find. On the whole, however, the relative abundance of meat was reflected in the importance the country placed upon maintaining farmland, as some estimates suggest the English had one of the highest ratios of domestic animals per capita. Yet as far as colonists in the New World were concerned, mere replication of English eating habits wasn't enough; all of that abundant, fertile land had to be used to produce *even more* meat.

The English established their first significant colony, Jamestown, in May 1607, where Williamsburg and the Chesapeake Bay are now. By this time, officials of joint-stock companies—colonies that were organized as a business venture funded by investors from the mother country—considered livestock to be a critical start-up cost associated with new colonies. The Virginia Company, for example, chartered by King James I to establish settlements along the coast, provided colonists with a regular supply of chickens, goats, pigs, and cattle. These animals did not come cheap, which may explain why the Pilgrims, who in 1620 arrived in Plymouth, on the coast of Massachusetts, did not bring any large livestock animals with them. Nevertheless, as one colonist wrote in a letter, by September 1623, "the town [had] six goats, about fifty hogs and pigs, also divers hens."[7] Though it is not known exactly when sheep first arrived in Plymouth, references to them begin to appear in the mid- to late 1620s.

Unlike the Pilgrims, the Massachusetts Bay Company decided livestock were to be a part of the colonist's settlement from the get-go. Under the leadership of John Winthrop, the colony's first governor, this second wave of Puritans sailed from England in April 1630 and established its center at Boston. Winthrop indicated there were 300 animals (240 cows and 60 horses) and 700 people on board. Over the next decade they were joined by tens of thousands of settlers and thousands more animals provided by the Massachusetts Bay Company and emigrating families.

How colonists interacted with their domesticated animals varied widely, depending on the colony. Consider colonists in the Chesapeake and New England colonies, who had very different approaches to farming animals. In the Chesapeake, colonists focused on raising pigs and cattle, in part because they required the least amount of attention and were well adapted to the new environment. (In contrast, sheep made for easy prey among wolves, and goats had a penchant for chewing the bark off apple trees, endangering the colonists' source of cider.) For the most part, domesticated pigs and cattle received minimal supervision, and some roamed not within the confines of penned fences and sheds but in the wild. The Chesapeake colonists largely allowed their animals to find their own food and shelter, no matter the season or weather. Sometimes, when the animals were left to fend for themselves, the Chesapeake colonists completely lost track of them. Though they marked newborn animals by branding their horns and clipping their ears in an attempt to protect their ownership rights, as they did in England, disputes inevitably arose both among the colonists themselves and with the natives.

Unlike the loosey-goosey farming practices of the Chesapeake colonists, New Englanders oversaw livestock in accordance with English tradition. New Englanders also raised pigs and cattle (though they placed greater emphasis on raising sheep for wool, despite the challenges with them), but they did so under a watchful eye, relying on wood-pasture farms, in which woodland provided shelter and forage for grazing animals, as well as sturdy fences that surrounded planting fields.

Though their approaches to farming were different, most English colonists marveled at the abundance of animals available to them, although it's hard to know whether this was an exaggerated ruse to encourage immigration. Francis Higginson, a Puritan minister, remarked in 1630 in a glowing report on Salem, Massachusetts, that "it is scarce to believe how our kine and goats, horses, and hogs do thrive and prosper here, and like well of this country."[8] In 1656 John Hammond, a resident of Virginia, wrote, "Cattle and Hogs are everywhere, which yeeld beef, veal, milk, butter, cheese, and made dishes, porke, bacon, and pigs, and that are as sweet and savoury meat as the world affords."[9] John Lawson, an English explorer who founded the two settlements of Bath and New Bern in present-day North Carolina, describes a land of plenty in his book *A New Voyage to Carolina:* "Their Stocks of Cattle are incredible, being from one to two thousand Head in one Man's Possession: These feed in the Savannas, and other Grounds, and need no Fodder in the Winter. Their Mutton and Veal is good, and their Pork is not inferior to any in America."[10]

At first, English colonists had no choice but to be largely self-sufficient. In addition to building their own homes and furniture, processing their own yarn to make cloth, producing their own soaps and candles, as well as hunting and fishing for food, they raised their own domesticated animals for meat. But it wasn't long before some settlers were able to step away from full-time subsistence and engage in varying degrees of trade and sale, including the purchase of meat from farmers who began specializing in producing a surplus. As tens of thousands of people immigrated to the colonies, towns formed with a diversity of artisans and merchants with particular skills, including butchery, who produced the goods and provided the services necessary for daily life.

Alongside this influx, important places and buildings of commerce were established. Many of these establishments were taverns (or the "ordinary" as it was referred to in Puritan Massachusetts), the first of which was opened by Samuel Cole in Boston on March 4, 1634. In addition to serving as a means of direction for travelers, as well as settings where they could eat, drink, be entertained, and spend the night, taverns were where farmers, artisans, and town merchants conducted business. At the center of many colonial towns was a market square, a large open space where farmers could come with their extra meat to trade or sell. Though these seventeenth-century commercial outlets paled in comparison to the well-established ones in England, they provided a meaningful forum of economic activity, including for the sale of meat.

In time, these intra-colony markets grew into burgeoning networks of exchanges that connected distinct colonies to one another. Farmers would stop seeing the towns in their immediate vicinity as the only markets and resources, instead turning their attention to distant places where better prices could be obtained. The earliest reference to meatpacking—a general term for the industry around processing and packaging livestock—as a commercial enterprise in America dates back to a pig slaughterhouse in Springfield, Massachusetts, established by William Pynchon in 1662 for the purpose of selling pork to the West Indies.

By the late 1680s, settlers in the Carolinas began to rely on slaves from the West Indies (who hailed from cattle-grazing regions in West Africa) to raise large herds of cattle for both local consumption and export. Soon after, the region was almost entirely responsible for provisioning the supply of meat in the West Indies. British records reveal that from 1768 to 1772, New England sent 239 dozen poultry (in addition to 13,496 barrels of fish) annually to the Middle Colonies—Pennsylvania, New York, New Jersey, and Delaware—in exchange for other goods. Pennsylvania and New York sent Virginia and Maryland 302 barrels of ham, and Pennsylvania and New York sent each other 21,944 barrels of beef and pork. These figures of inter-colonial trade of domesticated animals and meat throughout the Americas only scratch the surface of the actual market, as hardly any colony was completely self-sufficient and undoubtedly relied on this evolving web of trade routes to meet its needs.

English colonists also supplemented their diet of meat from domesticated animals with meat from an abundance of wild land animals, including rabbits, squirrels, possums, raccoons, deer, bears, beavers, moose, elk, swans, herons, eagles, and songbirds. Back home, with rare exceptions during dire times, hunting was treated not as an effort to achieve subsistence but as an act of leisure in game parks; proper meat came from domesticated animals. But throughout the colonies, hunting wild animals for meat was a sensible norm in a land teeming with roaming flesh. John Hammond, a resident of Virginia, wrote in 1656 that "wilde Turkeys are frequent, and so large that I have seen some weigh neer threescore pounds; other beasts there are whose flesh is wholsom and savourie, such are unknowne to us; and therefore I will not stuffe my book with superfluous, relation of their names."[11] Colonists also enjoyed harvesting the vast quantities of fish and shellfish in the North American waterways. The governing council of the colony in Virginia noted in its first letter home in 1607 that the waterways are "so stored with sturgion and other sweet fish as no man's fortune hath ever possessed the like."[12]

Perhaps no meat from wild animals was prized by colonists as much as venison from deer. In England, venison was regarded as a high-status food primarily eaten by the wealthy or during special occasions, but no such class distinctions existed in the colonies. The Huguenot visitor Durand of Dauphine wrote in 1687, "There are such great numbers of red & fallow deer that you cannot enter a house without being served venison. It is very good in pies, boiled or baked."[13] Some wild animals completely new to Europeans were so easy to hunt that they became endangered or extinct. Bison, which once spanned almost the entire United States, were reduced in population from tens of millions to mere hundreds; they have since made a comeback, in part due to conservation efforts. The passenger pigeon, which had a range extending from Canada and the Great Lakes to the Great Plains and south to Virginia, was not so lucky; armed with a wide variety of methods to capture and kill them, from nets to shotguns, settlers eventually drove them to extinction.

Whether from domesticated or wild animals, exactly how much meat did European colonists eat? With a lack of reliable data, nobody knows for certain. But the available evidence suggests they ate more meat in more varied forms than most people on the planet at the time. Estate inventories presented at probate court and widows' allowances as allocated in wills between 1620 and 1840 in Cambridge show an increase in meat consumption over this period. During the seventeenth century, only 31 percent of the households at the lower end of the wealth scale had stores of salt pork or beef. However, by the early eighteenth century, the regularity of storage of salted meats improved for all wealth classes. "By the time of the Revolution, domestic animals furnished most households in rural New England with a sufficient and nearly constant supply of meat,"[14] explains historian Sarah F. McMahon.

Anecdotal evidence from journals and letters during this period also points to an abundance of meat. For example, an astonished German visitor to Pennsylvania wrote in 1756 that "even in the humblest or poorest houses, no meals without a meat course."[15] Autobiographies tell a similar story. Susan Lesley, a memoirist who reminisced about her life in Northampton, Massachusetts, during the 1820s and 1830s, listed the meat dish first when describing her mother preparing dinner: "[There was] always a large joint, roasted or broiled, with plenty of vegetables and a few condiments . . . good bread and butter, and a plain pudding or pie."[16] Accordingly, meat occupies a large portion of the recipes in a prominent cookbook dating from 1700; in a cookbook written around 1740 by Jane Bolling Randolph, a descendant of Pocahontas and John Rolfe; and in the 1824 classic *The Virginia Housewife* by Mary Randolph. Meat is also heavily featured in *American Cookery* by Amelia Simmons, published in 1796 in Hartford, Connecticut, and *The Frugal Housewife*, published in Boston in 1829. These books give tips on judging the quality of meats at the market ("their smell denotes their goodness"[17]), and how to eat meat on a budget ("liver is usually much despised; but when well cooked, it is very palatable; and it is the cheapest of all animal food"[18]). Instead of traditional one-pot meals with meat, vegetables, and grains, the cookbooks suggest early American meals comprised many individual dishes, all anchored by a hunk of flesh.

This increased abundance of meat likely reflected an increase in fortunes among many of the colonists. By the mid-1700s, most of the colonies were quite prosperous. There was low unemployment, no income tax, and prices were generally stable. Commenting on this positive state of affairs, Benjamin Franklin wrote: "There was an abundance in the Colonies, and peace was reigning on every border. It was difficult, and even impossible, to find a happier and more prosperous nation on all the surface on the globe. Comfort was prevailing in every home."[19] (This was likely an exaggeration on Franklin's part, as others described much harsher realities for those who existed in lower socioeconomic circles.) For centuries, multiple meats had been the standard at European banquets. Now, colonists intent on maintaining their traditional fare continued to flaunt their affluence in part by the quantity and variety of meat they consumed at mealtimes.

Nevertheless, by today's standards, colonial methods of production and distribution of meat were still in their infancy. For the most part, the meat industry was highly localized; even by the turn of the nineteenth century, most Americans still lived in largely isolated agricultural households and small towns linked by horse-drawn wagons. Farmers either killed the animals in their backyards or had them butchered by local slaughterhouses, selling the fresh cuts of meat in marketplaces or preserving them via salting or smoking. But change was on the horizon, with burgeoning social trends and technological innovations ensuring meat consumption would continue to take center stage for generations to come.

4

Anywhere, Anytime, for Less

They use everything about the hog except the squeal. — *Upton Sinclair*

One night in 2017, I visited my parents in Staten Island for dinner. My mom passed me some mashed potato and green bean sides from Boston Market, though, despite my mom's efforts, the meal was a bit meager compared to her and my father's rotisserie chicken. As I regarded the glazed bird—which, admittedly, smelled delicious—I couldn't help think about how little it cost. Despite the work of hatching, feeding, housing, transporting, slaughtering, plucking, packaging, and shelving this two-pound chicken, it cost barely $8. In fact, for years now, Costco has notoriously priced its rotisserie chicken for a mere $4.99.

Meat has become so cheap in this country that your average family can consume a virtually unlimited supply of it for next to nothing. As we saw in the previous chapter, this is a distinctly modern phenomenon—and one that Americans all but regard as sacrosanct. But as I watched my parents pick apart their impossibly cheap chicken, I couldn't help but wonder: Was meat-eating more common now because the desire for it was stronger than ever before? Or was our ravenous appetite for meat always there, but only recently realized thanks to cheap prices? As I dove into the research, I discovered that this—ahem—chicken-or-the-egg dilemma has no easy answer.

In 1820, only about 7 percent of the nation's nine million citizens lived in a large town or city. By the 1860s, fueled by the technological and scientific developments of the Industrial Revolution, this figure had reached 25 percent. New York led the charge, reaching a population of more than one million in 1860 thanks to its ideal geography. In order to meet the carnivorous demand, livestock producers in New Jersey, Connecticut, Upstate New York, and Massachusetts worked with drovers who brought animals to convenient waterways and transported them by boats to the city. Many cattle ended up around the Collect Pond, a booming livestock district that occupied the space where Lower Manhattan's courthouses now stand.

Despite this rapidly increasing infrastructure, meat markets simply could not keep up with the demand of New York's skyrocketing population. By 1850, New York butchers were slaughtering nearly ten thousand animals per week to feed the urbanites. By the 1860s, the city's ravenous appetite increased to 1.1 million animals per year. Hard as they tried, farmers in the region's countryside simply could not raise animals fast enough. Weather was sometimes to blame, especially when droughts led to grain shortages, or when storms impeded transportation. When demand outstripped supply, meat prices soared, which likely explains why meat consumption declined from the early to mid-nineteenth century.

Help would come from farmers and entrepreneurs in the West in a series of spectacular innovations that transformed the meat industry. In the early nineteenth century, there was a vast migration toward the relatively unpopulated interior and valleys—a continuation of westward expansion and seizure of land from natives that had begun in the earliest days of American colonization. From 1800 to 1820, over half a million people moved into Ohio alone, while another six hundred fifty thousand settled in Indiana, Illinois, Missouri, Michigan, Alabama, Mississippi, and Louisiana. These regions offered enormous swaths of land for growing corn and raising livestock at a cost of just two dollars per acre—a twentieth of the cost of lands in the East—and ideally situated near rivers and canals for transportation to markets via the newly invented steamboat.

With its impressive network of waterways providing access to markets across the country, one of the most important pork producers was Cincinnati, Ohio. When hogs were ready to be butchered, farmers typically brought them to a slaughterhouse on the outskirts of the city or within the city itself in the central industrial district. To put it mildly, these droves were not easy. Based on the reminiscences of herders, a day's journey was five to ten miles, and depending on the distance, it could take weeks or months to arrive at the final destination. Herds of hundreds of hogs often required several drovers on foot and several on horseback to guide them into market. Hogs were also driven to slaughterhouses in other cities, but entrepreneurs in "The Queen City"—so called by Cincinnati locals in reference to the civilization the city offered in the midst of the wildness of the emerging West—quickly realized it was cheaper to drive and slaughter the hogs locally and to ship the processed pork on boats along the waterways than to drive the hogs to population centers hundreds of miles away. For this reason, it wasn't long before a seemingly endless number of hogs roamed the Cincinnati streets alongside its residents.

By the 1830s Cincinnati became known as "Porkopolis," and it began to attract much grimmer commentary: Isabella Lucy Bird wrote in her 1856 book *The Englishwoman in America*:

> The Queen City bears the less elegant name of Porkopolis; that swine, lean, gaunt, and vicious-looking, riot through her streets' and that, coming out of the most splendid stores, one stumbles over these disgusting intruders. Cincinnati is the city of pigs . . . by which . . . huge quantities of these useful animals are reared after harvest in the corn-fields of Ohio, and on the beech-mast and acorns of its gigantic forests. And at a particular time of the year they arrived by thousands—brought in droves and steamers to the number of 500,000—to meet their doom, when it is said that the Ohio runs red with blood![1]

In 1860, reporter Nicholas A. Woods vividly described Cincinnati's pigs for *Times of London* readers: "They come against you wherever you turn, from hogs, black, muddy unsightly monsters, down to little sucklings not much bigger than kittens, on which you inadvertently tread and stumble, amid shrill squeakings almost enough to blow you off your legs."[2]

Once slaughtered, "the hogs are taken into the pork houses from the wagons and piled up in rows as high as possible. . . . If the pork is intended to be shipped off in bulk, or for the smoke house, it is piled up in vast masses, [and] covered with fine salt . . . packed in barrels . . . while the residue is put up into prime pork or bacon . . . [and] is sent off in flat-boats for the coast trade." The quantity of barrels of pork and bacon were so large that "room cannot be found on the pork-house floors, extensive as they are, and which are, therefore, spread over the public landing, and block up every vacant space on the sidewalks, the public streets, and even adjacent lots otherwise vacant."[3] These meatpacking plants processed some eighty-five thousand hogs in 1833 in this way. By the 1850s, the city processed upward of half a million hogs a year. But as the *New York Times* noted on March 28, 1863, "The season of '61–'62 showed that Cincinnati was no longer entitled to the swinish distinction of being the chief pork-packing, city of the world, as Chicago then packed forty thousand more hogs than she did."[4]

Since its founding in 1833, Chicago had experienced some initial growth as part of the continuation of the westward migration, and due to its strategic location between the Great Lakes and the navigable system of waterways feeding into the Mississippi River. But Chicago's population would explode as a direct result of the city's proliferation of railroads. In 1850, nearly thirty thousand people lived in Chicago. By the end of the decade, this figure had almost quadrupled, and the city was on its way to becoming the nation's industrial capital.

Unfortunately for meatpackers in Cincinnati, the railroad did not offer the same level of explosive growth. One reason was their rivalries with other packing centers: Cincinnati had to deal with intense competition from other large cities, like Louisville, which also had rail links tapping into livestock production in the Ohio Valley. Also, many Cincinnati pork merchants were satisfied by the waterways that had worked for decades and reluctant to send pork directly east by rail. In contrast, Chicagoans were quicker to embrace innovation and exploit the improved commercial conditions brought about by railroads. With a boost from soldiers ravenous for meat during the Civil War, Chicago passed Cincinnati as "Hog Butcher for the World," producing a total of cured pork products valued at twice as much as the former Porkopolis produced.

Within just the first week of being open, Chicago's famed Union Stock Yards processed 17,764 hogs, 1,434 sheep, and 613 cattle. By 1870, they were processing 2 million animals every year; two decades later, they expanded that to 9 million. By 1900, it processed 82 percent of the meat consumed in the United States, and at its peak before closing in 1971, it processed over 18 million animals in a year.

Stockyards and packing plants like Union became morbid tourist attractions, as hundreds of thousands of visitors each year walked along-side uniformed tour guides who added commentary to the sights and smells of the full process—beginning with their ominous arrival on trains and ending with bloody loss of life on the killing floors. One of these tour-ists was Paul Bourget, an eminent French novelist and cultural critic who was drawn to Chicago because he had heard it best represented America, "with its contrasts of extreme refinement and primitive crudity."[5] On the tour, he watched a long line of hogs being driven up an incline chute to a small opening in the building. Upon entering the building, it gave off a vile stench that seized him "by the throat." He then walked across a sticky floor covered in a "sort of bloody mud," and made his way up several flights of stairs to the visitors' gallery. There, he watched the entire slaugh-tering operation, and later remarked this experience "will always remain to me one of the most singular memories of my journey."[6] English writer Rudyard Kipling reported having a similarly impressive experience, writ-ing, "You shall find them out six miles from the city; and once having seen them you will never forget the sight."[7]

Visitors to the United States continued to marvel at the amount of meat Americans consumed during the mid-nineteenth century. One visi-tor from England remarked in 1864, "As a flesh-consuming people, the Americans have no equal in the world. They usually have meat three times a day, and not a small quantity at each meal either. I have seen gentlemen choose as many as seven or eight different kinds of animal food from the bill of fare, and after having all arranged before him in a row, in the national little white dishes, commence at one end and eat his way through in half a dozen minutes."[8]

When the Union Stock Yards first opened, it processed mostly hogs and some cattle. But within just a few years, that ratio started to change. The influx of cattle was largely due to a migration to regions west of the Missouri River. When the Republic of Texas declared its independence from Mexico in 1836, there were only thirty-five thousand people living there. By 1860, the population had swelled to over six hundred thousand. The rise was in part due to "Manifest Destiny," the belief that US settlers were destined to charge westward and settle lands on their way to the coast. Among these settlers were farmers eager to take advantage of the superb grazing grounds and a species of wild cattle that roamed them. Texas Longhorn, a hybrid cross between southern cattle and wild black cattle first introduced by the Spanish, were teeming in the region. While the Civil War largely stymied the growth of the longhorn market, by the end of the war, as many as five million longhorns roamed free in Texas, a situation that exacerbated an already shattered economy. Due to the glut on the Texan market, longhorns were selling at less than $6 a head. As a result, ranchers looked eastward to cities like New York, where a head of cattle was worth upward of $80.

There were two big challenges in getting longhorn cattle from Texas to these markets. The first was that driving longhorns from Texas stockyards in Chicago was not easy. The feral cattle, with their famously large horns, were extremely temperamental, which made herding them difficult and dangerous. J. Frank Dobie, a folklorist who was born on a ranch in Texas in the late nineteenth century, affirmed the task as a challenging one:

> Ask any old-time range man of the south country to name the quickest animal he has ever known. He won't say a cutting horse, a polo pony, a wild cat, a striking rattlesnake. He doesn't know the duck hawk. He will say a Longhorn bull. . . . Some other bulls are quick; many breeds fight. . . . But none of them can bawl, bellow, mutter and rage like the bulls of Spanish breed and none can move with such swiftness.[9]

It took a team of about a dozen cowboys to drive several thousand of these unruly cattle over many months. Along the way, they endured brutal temperatures and constant bouts of torrential rain. As one cowboy remarked ruefully in his memoir, *A Texas Cowboy*, "Everything went on lovely with the exception of swimming swollen streams, fighting now and then among ourselves and a stampede every other stormy night."[10] Moreover, the cattle would lose so much weight during the journey that they would be hard to sell profitably upon arriving in Chicago.

The second challenge concerned a tick that infected cattle with "Texas fever," a disease characterized by unpleasant symptoms like high fever, emaciation, anemia, and bloody urine. Compounding the problem was that when tick-infested Texas Longhorn were driven out of the southwest, they mingled with local cattle and spread the disease. As a result, various states north of Texas enacted legislation establishing quarantine lines where no longhorn could pass. Though many cowboys ignored the law, locals would try to stop passing herds, sometimes resorting to beating or killing the drovers or stealing the cattle.

A successful Illinois livestock dealer named Joseph G. McCoy created a solution to both challenges in 1867. He devised a clever plan to build a southern stockyard near the Union Pacific Railway where longhorns could be driven permissibly and then shipped east for a large sum, thus avoiding the lengthy and expensive journey to Chicago. While McCoy was rebuffed in cities like Junction City, Solomon City, and Salina, he turned his attention to a "very small, dead place" in Texas called Abilene. First, he managed to buy a small parcel of land from a farmer who owned the town site. Then he convinced the state's governor to grant an exception in allowing him to build his stockyard and shipping facility even though it lay within the quarantine line imposed by the legislature. Lastly, he reached an agreement with the president of the Union Pacific Railway's Eastern Division, who pledged to provide recompense for every stock car of cattle shipped (a promise that would not be kept and marked the start of McCoy's rapid spiral into financial ruin).

Having secured all the agreements he needed, McCoy went to work and promoted an old trading route, a series of trails leading north from San Antonio known as the Chisholm Trail, to the railhead in Abilene. Cattle began to arrive there in August, and by early September the first shipment of animals left for Chicago. That first year, some thirty-five thousand head of cattle in Texas were driven to Abilene along that trail; the next year, seventy-five thousand; in 1870, three hundred thousand; and the following year, six hundred thousand. Over time, the longhorns were incrementally replaced by breeds that were less lean and produced meat faster, including shorthorns beginning in the 1850s and then black Angus and white-faced Herefords in the 1870s.

Although McCoy and his stockyard in Abilene would soon be outcompeted by entrepreneurs with more backing in nearby cities like Wichita, Caldwell, and Dodge City, he ushered in a new era of moving cattle in the Far West. By the 1870s, ranchers had driven cattle as far north as Idaho, Colorado, Wyoming, and Montana, where they began raising herds and driving them eastward to thriving meatpacking districts in the East including Cincinnati and Chicago, the latter of which became known as "the great bovine city of the world."[11] In 1909, for the first time, beef production surpassed pork production, and by 1952 some 219 million cattle had passed through their gates.

Despite all the progress made during the mid-nineteenth century in transporting both pigs and cattle from the West to eastern markets, the national meat industry was still a seasonal business, a major factor limiting its supply. For example, in Chicago's Union Stock Yards, during the period before the Civil War, the number of pigs slaughtered in July was a tenth less than the number slaughtered in December, due to a lack of refrigeration options during summer. Even as late as 1870, packers handled only 6 percent of the plant's annual volume between the spring and summer months. Although some hogs trickled into the stockyards during the off-season, they were slaughtered mainly for fresh meat that was consumed locally in Chicago. The seasonality of the cattle industry was even more problematic. While most urbanites were content eating preserved pork in one form or another, they were much less willing to accept preserved beef; due to its tough, fibrous nature, it tasted lousy compared to preserved pork. For that reason, many cattle were shipped to a local processing facility where they were slaughtered by wholesalers and delivered fresh to nearby butcher shops for immediate retail sale.

While a lucrative business overall, costly inefficiencies were inherent in transporting live animals. For starters, livestock were fed and watered at terminal facilities along the route, but that didn't prevent many of them from losing weight (or even dying) on the lengthy journey. "The poor beasts are herded close together, cramped, bruised and lean," one writer explained in an article in *Scribner's*. "In a journey of over a thousand miles, occupying five or six days, they have been fed and watered only twice."[12] As an early twentieth-century writer in New York's *Butchers' Advocate and Market Journal* explained, "It is evident that an undue amount of time required for the transportation of livestock does result in a loss of weight which, in turn, reduces the amount of money to the shipper and the amount of meat available for the consumer."[13]

The only way to eliminate this problem was to invent some sort of cooling apparatus that would allow meatpackers to send butchered meat from one place to another. The idea was not a new one. In 1842, in one of the first references to the subject, *The Boston Traveler* printed the click-bait-worthy headline "Freaks in Railroad Transportation." The article described a world filled with "refrigerator cars, in which fresh beef, pork, veal, poultry, pigeons, venison, wild game, and other fresh meat can by a moderate quantity of ice, be kept in perfect order in the heat of summer."[14] But when American railroads debuted this icebox on rails, they moved milk and butter only. As early as 1843, dairy farmers in Orange County, New York, sent milk on the Erie Railroad in cans containing a tin tube filled with ice that kept the milk cool until arrival four-and-a-half hours later in New York City. When the Rutland Railroad debuted a boxcar containing blocks of ice and heavy-duty insulation on July 1, 1851, it carried eight tons of butter from New York to Boston. However, these methods could not stay cold enough to safely carry meat from place to place, so the hunt for a better solution continued.

One proposed solution came from J. B. Sutherland of Detroit, Michigan, in 1867. His design included large tanks of ice at each end of the boxcar with vents above the tanks that provided air circulation to the interior. But riddled with technical challenges and an inability to raise funding, Sutherland abandoned the idea. One year later, a Detroit merchant, William Davis, patented an improved design using metal racks that suspended the carcasses above a frozen mixture of ice and salt. Impressed, a banker named Caleb Ives envisioned massive profits using Davis's refrigerated cars to ship dressed beef to eastern cities. Joining forces with a local butcher, George Hammond, Ives built a refrigerated car according to Davis's specifications and assumed the risks associated for the first shipment while Hammond selected the cattle, butchered and packed the beef, and sent a shipment to Boston. In the spring of 1869, the meat arrived safely, and the partners considered the experiment a success. Hammond purchased the patent from Davis and later cofounded the Dressed Beef Transportation Company. The team established operations south of Chicago's Union Stockyards in a remote and unpopulated location in northern Indiana that offered rail connections with eastern cities as well as an inexpensive supply of ice from nearby lakes. By 1875, the company sold $2 million worth of dressed meat per year.

Despite successfully shipping some fresh meat from the West to the East via refrigerated cars, there were still many kinks to work out, and each was an obstacle to widespread adoption of refrigeration. The first limiting factor was the technology itself. For starters, in early designs, the carcasses frequently came into direct contact with ice, which blackened the surface of the meat—which consumers and public health advocates interpreted as a sign it was unfit for consumption. To fix this problem, the carcasses were suspended from the ceiling. Unfortunately, this remedy had tragic side effects: As the train zoomed around a curve, the immense momentum of swinging meat could cause derailments. But the primary obstacle was that pesky tendency of ice to melt, and a basic system for refilling the ice bunkers at stops along the way simply did not exist. (Although ice-making machines had become available in the United States during the 1870s, they were too heavy and bulky to be mounted in railroad cars.)

Real progress in improving refrigeration technology and building the necessary infrastructure to capitalize on it was made in the mid-1870s with a man named Gustavus F. Swift. Swift grew up in West Sandwich, Massachusetts, on his father's farm. After working alongside the town butcher for years, he opened his own butcher shop in Barnstable, where he engaged in the business of buying and selling livestock, and quickly became known as one of the best traders in New England. To capitalize on his glowing reputation, Swift entered into a partnership with a renowned Boston livestock buyer, James A. Hathaway, who was shipping cattle to markets in England. The partnership became official in 1872 under the name Hathaway and Swift. Three year later, when Swift became aware of the tremendous opportunities in the West, he moved the business's headquarters to Chicago where he bought large numbers of livestock for shipment by rail from Chicago to Boston.

In the late 1870s, Swift realized that he could save a lot of money on freight charges by having his animals slaughtered in the Union Stock-yards and shipping only the dressed—or partially butchered—meat to the eastern markets. He began experimenting with sending fresh beef east on trains in the winter, leaving the freight-car doors open to keep it cold. Dissatisfied with only seasonal shipments, he hired Boston engineer Andrew Chase to design a ventilated car that was better insulated than Hammond's. Chase placed the ice bunkers in the top corners of the car, allowing the chilled air to flow naturally downward and the warm air to escape through ventilators, while the ice bunkers could be easily replenished with ice from the outside, making the car more efficient to operate. Swift approached numerous railroad companies and requested they build and install refrigerated cars according to Chase's schematics. But the railroads, protective of their expensive and highly profitable livestock cars, weren't interested. But Swift did not give up, and eventually he struck a deal with Grand Trunk Line, a railway operating system that at the time had little livestock interest due to its somewhat longer, circuitous route, and was eager to make headway into the meat industry. If Swift could supply the refrigerator cars and construct re-icing facilities along the route, Grand Trunk Line was willing to carry dressed meat.

Swift went to work, investing his own capital into ten refrigerator cars built by the Michigan Car Company and established an integrated system for operating and re-icing them. Rather than relying on wholesale distributors—whose job it was to solicit orders from retailers in surrounding towns and villages—Swift recruited new ones. Using the telegraph, orders placed by these distributors specifying the breeds, grades, and quantities were relayed during the day to headquarters and to buyers at the stockyards. Livestock arrived at the stockyards at night, were purchased in the morning, and slaughtered by the early afternoon. The magazine *Harper's Weekly* marveled that once processed and loaded onto the refrigerator cars, "the *meat* is *easily transferred* to the *storage room*, which is of the *same temperature* as the *car*, *without loss* of *time* and *without being removed* from the *hook* on which it was hung when killed." These storage rooms, also known as "branch-houses," were the crown jewel in Swift's masterful distribution network, providing local butchers with easy access to supplies of his dressed meat.

Though a highly risky venture for Grand Trunk Line and Swift, it worked out brilliantly for both. By 1879, Swift had created the first refrigerated train service delivering dressed meat long distances year-round from Chicago to markets in New England. He was so successful that by 1884, New England's live cattle shipments had almost entirely ceased.

By the time Swift's dressed meat arrived in New York City, he wasn't the only game in town. Among those now shipping dressed meat in refrigerator cars from the West to cities in the East was another Chicago meatpacker named Philip Armour. Armour was born on a farm in Stockbridge, New York, in 1832. In 1852 he recruited a neighbor and walked across the country to California with only some clothing, a pair of boots, and dreams of prospecting gold. Upon arriving, Armour quickly realized his natural talent was not panning for ore but scoping out business opportunities during the gold rush. He purchased an aqueduct that moved water to diggers and washers, and when that proved profitable, he purchased several more. Eventually he sold them all in favor of a new opportunity: moving provisions from the West to the East.

Using the same business acumen he had fostered in California, he and a few partners made a small fortune in the grain and pork businesses in Milwaukee. In 1867 he rented a small plant on the South Branch of the Chicago River and began packing hogs. One year later, Armour and his younger brothers founded Armour and Company in Chicago and purchased a larger plant to scale their business to cattle and sheep, and by 1871 their company was the fifth-largest of the city's twenty-six packers. That same year, Armour purchased land near the Union Stockyards to build an immense meatpacking plant. In order to get his meat products to market, he followed the lead of his rival Gustavus Swift by establishing the Armour Refrigerator Line. By 1900 Armour and Company had twelve thousand yellow train car units in its roster (one-third of all the privately owned cars in the country), all built in Armour's own car plant. "Armour yellow" became the unofficial color of refrigerated train cars, and each of Armour and Company's yellow cars bore the slogan: "We Feed the World."

It was in this climate of competition that Swift, Armour, and other leading packers began revolutionizing the mechanization of meat processing. Cattle and hogs were processed or "disassembled" with unprecedented efficiency, with each meat-processing step reduced to its most basic component. In fact, the idea of the assembly line to build cars came to Henry Ford when he saw the disassembly lines used to process meat at the Armour and Swift meatpacking plants. Innovative technology included overhead trolleys that transported carcasses and automated conveyer belts that whisked meat through the plant. Along the way, each employee specialized in a specific task and repeated the same motion over and over again to minimize time wasted. At their height, some of the kill floors employed tens of thousands of workers who processed hundreds of animals per hour. And rather than toss away the "inedible" parts, every part of the animal carcass was used to maximize a profit in a burgeoning by-product industry. Dried blood, for example, was used to make fertilizer, and skin and thigh bones were used to make knives, brush handles, and buttons. As one comedic commentator named Finley Peter Dunne put it, "A cow goes lowin' softly in . . . an' comes out glue, gelatine, fertylizer, celooloid, joolry, sofy cushions, hair restorer, washin' sody, littrachoor an' bed springs so quick that while aft she's still cow, for'ard she may be anything fr'm buttons to pannyma hats."[15] Thanks to this ruthless, "waste not," profit-seeking efficiency and centralized ownership of a widespread distribution system, more Americans now ate more meat than ever before. As fresh beef from the West flooded eastern markets near the turn of the twentieth century, prices plummeted by as much as 30 percent.

Somewhat paradoxically, consumption of red meat soon fell as quickly as it had risen. There were at least four major reasons. First, due to additional population growth, output of meat once again could not keep up with demand, and prices often skyrocketed. In 1899 meat accounted for 28 percent of the total manufactured-food output in the United States; by 1937, this figure had dropped to 13 percent.

Second, there were heightened safety concerns about the quality of meat following the publication of *The Jungle* by Upton Sinclair in 1906. Sinclair had hoped to shock readers with his heroic account of "the inferno of exploitation"[16]—the ghastly treatment of immigrant workers in Chicago's slaughterhouses. Instead, *The Jungle* created a public outcry over the unsanitary conditions it described—how moldy meat was repurposed after being dosed with borax and glycerine, how meat tumbled onto filthy floors infested with rats, and how tripe and cartilage were dyed and flavored with spices so they could be sold as delivered ham.

The third reason why meat consumption declined was World War I. By the time the United States entered the war, countless citizens of allied countries were starving. The farms of Western Europe were carved into battlefield trenches and the farmers themselves became soldiers. In an effort to feed struggling allies and US soldiers overseas, President Woodrow Wilson created the US Food Administration to manage the wartime supply, conservation, distribution, and transportation of food. One of the administration's first campaigns asked US citizens to cut back on meat to help feed their four million soldiers stationed overseas. To spread the word and galvanize support, the administration disseminated posters, articles, pamphlets, and other educational materials, many sporting the slogan "Food Will Win the War." The campaign was a massive success: The United States shipped 18,500,000 tons of food to troops and allies in Europe between 1918 and 1919 while reducing consumption at home by 15 percent during the same period.

Finally, during the Great Depression, Americans simply struggled to put food on the table. By the start of 1933, unemployment had reached nearly 25 percent. With no welfare system in place, hundreds of bread-lines in New York City dispensed roughly eighty-five thousand meals a day to the hungry. It wasn't until the mid-twentieth century that another set of historical circumstances and innovations would hit the industrial meat scene, one that would both give beef a boost, but more significantly, take poultry on a meteoric rise to become the most popular dish in the United States.

5

From Farm to Factory

If slaughterhouses had glass walls everyone would be vegetarian.
—*Paul McCartney*

In the summer of 2018, I visited the Vernon, California–based Farmer John pork-packing plant, part of the Clougherty Packing Company, a subsidiary of Smithfield Foods. I arrived in what felt like the middle of nowhere outside the nine-hundred-thousand-square-foot facility at around 2 a.m. to do something very far outside my comfort zone: to give last sips of water to hogs being sent to slaughter.

I had been invited by Anita Krajnc, the cofounder of The Save Movement, to attend the "pig vigil" alongside hundreds of hardcore animal activists. Krajnc made headlines in 2015 when she was arrested and charged with criminal mischief for giving water to pigs on a truck headed for slaughter in Burlington, Ontario, Canada—what she describes to me as "an act of mercy"[1]—only to be arrested again in 2016 for similar actions. Up until this moment, I had never participated in a protest or vigil, let alone one for farmed animals. As I stood apprehensively among the crowd waiting for the hogs to arrive, a police officer began to speak. He explained that he and other members of the Vernon police department were there to oversee the proceedings, then itemized numerous ground rules to keep everyone safe, including that we must stay on the sidewalk until the truck bearing the pigs arrived.

Around 2:30 a.m., an enormous stainless-steel vehicle with air holes punched on the sides lumbered toward us. Just before stopping about twenty feet away from me, my senses were assaulted: I heard muffled snorts and high-pitched squeals and smelled manure so pungent it hurt the back of my throat. As I struggled to take it all in, I watched tear-stricken activists spray water from hoses into the mouths of the thirsty hogs. The chaotic and grim scene felt like a *Black Mirror* episode, only stranger and real. Shortly after the truck departed, the next one arrived, and this time I approached to get a closer look. What happened next surprised me: I made direct eye contact with one of the hogs, and I was instantly struck by how similar he was to my dog, Tobey. Suddenly I found myself sobbing uncontrollably. Several of the activists wrapped me in a hug.

By 3:15 a.m., I felt I had experienced more than enough and was ready to go home. As I made my way back to the car, I noticed along the exterior wall of the meatpacking facility a bright, colorful mural depicting happy pigs roaming free on pastures. The Orwellian image transformed my unprocessed grief into a ball of outrage because I knew it was a blatant lie: Not only had these pigs suffered in the hours leading up to their end, but they had also spent all of their preceding moments enduring the horrors of a factory farm. An hour later, I was finally in my bed, and as I lay in the dark trying to make sense of what I saw, I only knew I would never forget it.

It took me several days to recover, but as the time passed, I became grateful for the experience. Yes, the memory of it was painful, but it was also eye-opening. My initial objection to factory farming was based more on an abstract consideration of environmental harm and the suffering animals must endure. This experience put a face to that suffering. Those pigs' terrified squeals will reverberate in my mind every time I utter the phrase "factory farm." Still, the experience left me with a burning question: When and why did factory farming emerge?

The first recorded use of the term "factory farming" appeared in an American journal of economics in 1890, decades before the emergence of industrial animal farming. Alfred Marshall, a British economist, proposed the development of "factory farms" organized around the principles of factory management. He projected the farm machinery would be "special-ized and economized, waste of material would be avoided, by-products would be utilized, and above all the best skill and managing power would be employed."[2] Wilbur Olin Atwater, an American chemist who laid the groundwork for the modern science of nutrition, shared a similar senti-ment in *Century Magazine* one year later:

> The agriculture of the future will perhaps be a manufacturing process with correspondingly increased product. It may seem paradoxical to say the dense population which the older economy told us is to be the precursor of starvation, will be actually the antecedent condition of a cheap and abun-dant food supply; but is this anything more really than that re-assertion of a principle which has proven itself true in the manufacture of cloth in the factory, of machine in the machine shop and in countless other ways?[3]

While urbanization and the industrial revolution had certainly laid the groundwork, it was primarily in the 1920s when these early concepts of factory farms were realized. They emerged not with pigs or cows but with chickens through a dizzying array of transformations impacting nearly every aspect of animal agriculture.

This might seem paradoxical considering that until World War I, chickens were, for the most part, considered undesirable. In colonial times, unlike other domesticated animals, chickens were simply listed as a number on wills, inventories, and property statements; often, they weren't recorded at all. Caring for chickens consisted of merely allowing them to roam the farmyard for bugs and worms and tossing them excess grain and table scraps. The apathy toward them was largely a product of simple economics: Farmers could make more money on the eggs hens laid than from the skimpy flesh on their bones—unlike cows and pigs, egg-laying hens were worth more alive than dead.

On American farms, it was traditionally the job of women to raise a small flock of chickens and of the children to collect the eggs. This gender- and age-based sentiment was reflected in an 1894 edition of the *Annual, Issue 7* by Minnesota Farmers' Institute, which explained the practice of raising chickens was "entirely beneath the dignity of a full-grown man ... [and therefore] they would not spend their time fussing with a lot of hens."[4] Some of the eggs were eaten by the family and any surplus was sold locally. But eggs were not massive profit-generators, as typical pre-twentieth-century hens laid barely two dozen eggs per year. And since chickens were inexpensive to keep alive for egg production, the slaughter and distribution of their meat was limited, driving up the price. In time, chicken was even considered to be a luxury, hence the origin of the phrase a "chicken in every pot," touted by the Republican Party as a promise during the 1928 presidential campaign.

One of the first people to transform chicken production from a millennial-like side-hustle in the "gig economy" into something more financially substantial was Cecile A. Steele. And like many pivotal moments in history, it occurred entirely by accident. According to what has become an industry legend, Steele was an egg producer on the Delmarva Peninsula, a two-hundred-mile finger of flat farm country east of the Chesapeake Bay. In the spring of 1923, she ordered 50 laying-hen chicks from a hatchery a few miles down the road. The company accidently sent her ten times what she had ordered: 500 chicks instead of 50. Rather than send them back, Steele decided to make do with them all. She asked a lumberman to build her a small shed to house them and attempted to raise and sell them for meat. After eighteen weeks, she sold the 387 that survived for sixty-two cents a pound—several times more than what she would have made by selling their eggs over several years. Despite the chickens' small size, chefs and homemakers appreciated their versatility—they could be fried, broiled, roasted, or stewed—and wealthy consumers liked their taste. The next year, Steele ordered twice as many. The venture was so profitable that her husband left his job with the Coast Guard to help her raise chickens. By 1926, they were growing 10,000 birds at a time. Word spread quickly and others began to raise flocks for the explicit purpose of selling broiler meat. (Broiler is a term used to describe any chicken bred and raised specifically for meat production.) Less than a decade later there were at least five hundred farms like Steele's, and the Peninsula as a whole was producing 7 million broiler chickens per year.

To get these broilers to market, buyers would come to farms with a "catching crew"[5] late at night while the broilers were sleeping, then herd the birds toward one end of the chicken house where they were easier to catch. They were then placed into small cages, loaded onto the back of trucks, and driven to the dealer's holding station or assembly plant. Upon arrival they were transferred to new cages and fed overnight in preparation for transit in the morning to niche urban markets on the East Coast.

New York was the most popular destination for broilers, largely because the city was home to nearly two million immigrant Jews who had come to the shores of Ellis Island in the nineteenth century to flee anti-Semitism in Eastern Europe. Chicken could be slaughtered in front of Jewish customers to ensure it met kosher requirements. Jews believed a holy day was particularly honored if a luxury meat was served; because chicken was several times more expensive than other meats, it fit the bill as an ideal entree for the Sabbath and other Jewish holiday meals. In time, demographics of chicken consumers would diversify, but the initial success of farmers in the Delmarva Peninsula was primarily due to Jewish consumption practices. Indeed, a study conducted by the US Department of Agriculture in 1926 found that Jews accounted for as much as 80 percent of the live poultry sales in New York City.

During this time, hundreds of companies began developing increasingly specialized feeds for chickens of these expanding numbers. In the early 1900s, in part through a series of cruel experiments in which beagle puppies were housed indoors and fed a diet of only oatmeal, scientists discovered vitamin D was critical for the prevention of deficiency diseases such as rickets. Around the same time, using similar experimental methods on rats, scientists discovered vitamin A, important for normal vision, the immune system, and reproduction. The implications for animal agriculture were quickly realized, as chickens too suffered and fell ill when deficient in these vitamins. Adding these nutrients to feed meant chickens could survive without sunlight (the primary source of vitamin D) and without green grass (a source of vitamin A).

Consequently, chickens could be raised in large, roofed sheds with temperature, diet, and lighting controlled for maximum weight gain, all while reducing incidences of injury from inclement weather and predation from rats, foxes, skunks, raccoons, and hawks. On the whole, this made farming poultry much easier and less expensive. In 1923, it took 4.7 pounds of feed to produce one pound of broiler meat; by 1941, only 4.2 pounds of feed was needed to get that same amount of meat. Given that chicken feed represented between 50 to 60 percent of the cost of raising broilers during the 1930s, this 10 percent reduction amounted to huge savings. In addition to requiring less feed, broilers in intense confinement also grew faster and heavier. In 1927, the average broiler in Delmarva went to market after four months and weighed 2.5 pounds. In 1941, it was sold after only twelve weeks and weighed 2.9 pounds.

Although Cecile Steele and the farmers she inspired would be among the first and most influential manifestations of agribusiness, chicken consumption was still relatively low; as late as 1940, the average American ate only 10 pounds of poultry annually. Even Armour and Swift, who had invested in poultry operations in the Midwest in the early twentieth century, found it difficult to achieve economies of scale. While less expensive than it had historically been, chicken still remained approximately two and a half times as expensive as red meat.

Nevertheless, demand was climbing steadily. A new generation of younger Jews in the late 1930s was content eating chicken that had been slaughtered and processed near the chicken-growing areas in a "New York Dressed" fashion—that is, with the feathers plucked and the blood drained but with the head and feet still on the bird and all the insides still present—then packed on ice and trucked to market. As was the case with beef and pork, shipping New York Dressed chickens instead of live chickens offered considerable savings. Driven by consumer acceptance of New York Dressed, several local processing plants opened in Delmarva in the late 1930s. By 1942, ten had the capacity to slaughter, dress, and icepack a combined total of thirty-eight million broilers per year. Washington, D.C., received 6 percent, 10 percent went to Philadelphia, and a whopping 77 percent went to New York City.

By the time the United States entered World War II, the Delmarva Peninsula was overwhelmingly dominant in the breeding of broiler chickens, producing ninety million annually (more than half the country's broilers), up from seven million just eight years prior. Unfortunately for the Delmarva, its tremendous success was its undoing. In a dramatic turn of events, in 1942, the War Food Administration (WFA)—the agency responsible for the administration of the US Army overseas and allies' food reserves—commandeered Delmarva's entire poultry production operation, quite literally cutting off all major roads and seizing chickens from trucks headed in and out of the area. As one reporter put it in a *Life* magazine article of April 5, 1943, "History proves that when nations go to the war the one which fails to provide its people and its fighting men with ample food is the one that crumbles."

In service of this belief, one of WFA's then rather direct slogans was "Chicken is for fighters first!"[6] And indeed the Allies needed a lot of it: US farmers were asked to produce a staggering 4 billion pounds of dressed chicken, and the poultry industry in Delmarva would play a pivotal role. With the area's entire supply of chickens seized, non-military customers had to look elsewhere for poultry. The southeastern states, which had infertile terrain less suitable for other farming purposes and an ample supply of struggling family farmers, would go on to fill that void, becoming the new center of the industry.

The person most commonly credited for first scaling poultry production in this region was Jesse Dixon Jewell. Jewell's early life resembled something out of *A Series of Unfortunate Events* by Lemony Snicket. Jewell was born in Gainesville, Georgia, in 1902. His father, Edgar Herman Jewell, owned a prosperous cottonseed mill and was among the first in Gainesville to own an automobile. But then the first of many tragedies struck: Edgar died when Jewell was only seven years old. After attending the University of Alabama and Georgia Tech, he returned to join his mother and stepfather to work in the family business. Then, just a few years later, in 1930, his stepfather died. With his mother heartbroken and exhausted, and his wife and daughters financially dependent on him, Jewell had little choice but to step up and take over the family business.

Like most farmers in Georgia during the late 1920s and the 1930s, Jewell and his family struggled to pay the bills. The life of a farmer in the South had never been easy, but the economy worsened during the Great Depression in part because of a decade-old farm crisis preceding it. During World War I, farmers had ramped up production to meet the needs of US soldiers and war-ravaged Europe. But when demand plummeted after the war ended, farmers had little choice but to dump crops and livestock on the market at fire-sale prices. By 1933, corn was down to 19.4 cents per bushel from its prewar average of 64.2 cents; wheat was down to 32.3 cents per bushel from 88.4 cents; cotton was down from 12.4 cents per pound to 5.5 cents; and hogs were down from 7.2 cents per pound to 2.9 cents. By Jewell's own estimation, the family business was "shot."[7] Amid all of this, yet another tragedy struck: In 1936, not one but two deadly tornadoes hit Gainesville, killing more than two hundred people and damaging or destroying many more hundreds of homes.

As a matter of survival, Jewell searched for a way to supplement his income and found an opportunity buying chickens from local farms and reselling them in Atlanta and other urban markets. It was a challenging way to make what only amounted to some extra pocket change. Jewell's financial position was so dire that he had to immediately sell what he had bought from the farmers in order to cover the checks he had written to bankroll their purchase in the first place.

But Jewell knew demand for poultry was high, and he had a clever idea for how to capitalize on it, one that drew on an existing credit-based system that had dominated the South since 1865. Before the American Civil War, small-scale Southern farmers were generally self-sufficient, capable of feeding themselves and their families without relying on local stores for food supplies. But the war ravaged the South, and veterans who returned home saw devastation everywhere, especially on farms and plantations. Farmers were forced to rely increasingly on credit from storekeepers like Jewell—many of whom had relationships with wholesalers in the North— for the supplies needed to plant and grow their crops. But the credit came with two big stipulations. The first was that in exchange for the seed, feed, fertilizer, and other necessities in advance of the harvesting season, storekeepers demanded farmers specifically grow cotton. The reasoning was simple: Cotton was a safer bet than other crops. Unlike more perishable farm products, it was able to withstand the abuse of being transported over dirt roads and in derelict boats, and it was less bulky than other crops and therefore easier to ship in bulk.

Second, storekeepers required farmers to accept a "lien" on the cotton—essentially a mortgage-like right to keep possession of it until the debt owed was paid off. In order to secure the lien, farmers had to agree to pay a premium price for goods bought on credit as well as high interest rates on the loan—as much as 15 percent—along with other extraneous fees. Once the harvest season came, the farmers sold their crop, paid back the storekeeper, and kept whatever money was left over. This was usually very little, and sometimes farmers didn't sell enough crops to pay back the storekeeper and were forced to begin the following year in the red.

Naturally, many farmers in Georgia resented the crop-lien system, but few had any other choice. Jewell's ingenious idea was simple but revolutionary: Why not adapt and apply this detested but effective system to poultry production?

To realize his idea, Jewell first persuaded local destitute farmers to switch from growing crops like cotton to raising chickens. Then he bought one-day-old chicks from a hatchery and supplied them, along with chicken feed from his own business, to the farmers on credit. He also offered cash loans so they could convert their barns into chicken houses. Then, once the chicks became fully grown broilers, he packed them in his truck and drove to Miami, Florida, where he sold them at local markets. The system generally worked—farmers were usually able to repay their debts to him and earn a modest profit, while Jewell kept the lion's share and was able to save his family business from bankruptcy. Though the good times wouldn't necessarily last, many of the farmers in his expansive network were financially successful, most of them for the first time in their lives. But Jewell was just getting started. He would soon grow his business into the largest integrated chicken producer not only in the United States, but in the world.

By the late 1930s, Jewell had contracts with a small army of chicken farmers, but there was a problem: He was having trouble securing enough chicks from hatcheries to keep up with the demand. So, Jewell decided to build his own. Just before World War II he opened his own processing plant to slaughter and package the chickens his contracted farmers raised. Although chickens had previously been New York Dressed, Jewell found consumers outside of the New York area were more than willing to buy processed birds with the head, feet, and entrails removed. So, his birds were packed in ice and shipped to the marketplace in wooden barrels and marketed as "pan ready."[8] Production continued to skyrocket: In 1934, Georgia produced only four hundred thousand broiler chickens; by 1942 it produced ten million. By 1945, almost thirty million. The established nature of the poultry industry in Georgia was reflected in a colorful story about a boy who was hauled before the judge for making moonshine. He had previously been arrested for the same crime and received no more than a slap on the wrist, but this time he received a large fine and a jail sentence. When he protested that the court had never treated him this harshly before, the judge replied "I know, but now a man can make an honest living in these parts by raising broilers, so he doesn't need to break the law by moonshining any longer."[9]

Two occurrences gave Jewell's operation a significant boost during the 1940s. The first was World War II. As devastating as it was to farmers in the Delmarva Peninsula that the government seized their poultry farming operations, it tremendously benefited Georgia. Outside of the Delmarva Peninsula, chicken was available on the free market without government control. And unlike pork and beef, which were almost entirely earmarked for the troops, poultry was not rationed. In fact, the USDA actively encouraged citizens to eat more chicken through its "Grow More Poultry" program, touting its consumption as a form of patriotic duty because it freed up other meats for soldiers. All in all, by the end of World War II, chicken consumption had increased 50 percent nationwide.

The second reason for Jewell's success was a combination of government money, funded research, and intellectual capital. As part of an effort to restore prosperity to Americans through The New Deal, the federal government began tinkering with the broader agricultural system. In 1935, for example, the USDA launched the National Poultry Improvement Plan "to provide a cooperative industry, state, and federal program through which new diagnostic technology can be effectively applied to the improvement of poultry and poultry products throughout the country."[10] In 1936, the government encouraged the spread of electrical distribution systems in rural areas by offering loans for their installation. While 90 percent of those living in urban areas had electricity by the 1930s, only 10 percent of rural dwellers and farms did. With electricity, farmers were now able to raise chickens in confinement year-round, no longer limited to the sale of "spring chickens." Devices like electric lights, waterers, and heaters saved farmers time and money, allowing them to increase the number of houses and the number of chickens in their flocks.

After World War II, Jewell began scaling his operation. By 1954, he added the final touches to his operation: a feed mill to process grain shipped in from the Midwest, and his own rendering plant to process animal by-products. Jewell now had near complete control of growing and selling poultry, from incubation and hatching to processing, storage, and distribution. The exception was the broiler house, where chickens were raised until they were ready for slaughter. Jewell chose not to invest in broiler houses because he felt they were the riskiest and most volatile part of the business. Disease, for example, could and did kill off entire flocks of fifty thousand chickens. Instead, Jewell passed the risk onto the farmers, who took on a treacherous amount of debt to build broiler houses. This quasi "vertical integration"—the consolidation into one company of two or more stages of production normally operated by separate entities—set the standard for poultry processors everywhere, as dozens of other so-called integrators copied Jewell's design, not just within Georgia, but in the Delmarva Peninsula, Arkansas, North Carolina, Alabama, Mississippi, and many other states.

Two of these integrators following in the footsteps of Jewell were Arthur W. Perdue and John W. Tyson. Perdue started his chicken business in 1917 in Salisbury, Maryland, with the sale of eggs and laying hens, but by the early 1940s he began to switch to broilers when he realized the future was in their meat. Like Jewell, he never owned a broiler house and relied on contracted farmers to raise his chicks. By 1950 he owned hatcheries, feed mills, and slaughterhouses, incrementally consolidating the means of poultry production. Tyson began as a middleman transporting chickens in the Southern states in the early 1930s. Like Jewell and Perdue, whenever he came across a glitch in the system, he simply fixed it himself. When chicks and feed were in short supply, for example, he launched his own hatchery and feed mill. Although he started out by raising most of his own birds, he too began to embrace contract farming by the late 1940s. By 1958 he built his own processing plant in Springdale, Arkansas, making his company one of the first integrated broiler businesses in the state.

Businessmen like Jewell, Perdue, and Tyson embraced vertical integration for one obvious reason: It was highly profitable. Among other practical benefits, vertical integration eliminated, or at least greatly reduced, the transaction and handling costs associated with separate groups overseeing the various production steps. What Armour, Swift, and other meatpackers did to slaughterhouses—structuring them as factories and incorporating them into complex distribution systems across the nation—these integrators did for the entire poultry industry, building tight-knit corporations connecting every single aspect of the farming process.

While vertical integration helped meet demand for chicken and ease the burden of shortages in labor during the war, the poultry industry faced another major challenge: There simply wasn't enough feed to grow the animals, and the feed that was available was expensive. Many farmers couldn't afford the recommended feed with vitamin-rich fishmeal or cod liver oil, so they switched to entirely plant-based feed. Devoid of essential nutrients, the chickens grew slowly and only reached a relatively small size. Increased production costs coupled with less output of meat translated into high prices, limiting consumption. The situation for chicken producers only worsened when demand for poultry decreased after the war, with red meat once again available for citizens. The federal government canceled contracts, and producers now had a huge surplus of chickens and an inability to cut their feed costs. From 1942 to 1948, despite sinking demand, the price for broiler chickens increased 57 percent. If industry couldn't find a suitable replacement for its conventional feed, poultry consumption would plateau, or even revert back to its occasional Sunday dinner status.

It wasn't for lack of trying. Though scientists had been successful in determining the vital role vitamins A and D played in chicken development, they knew there was something in traditional animal-based feed that plant-based alternatives lacked. This was especially strange because scientists were able to obtain these mysterious growth-promoting, disease-resisting elements from some bacteria that lived on vegetable foods—and yet, they didn't seem to be in the vegetables themselves. With feed shortages during the war intensifying the need for answers, scientists began hunting for the so-called animal protein factor (APF) that had eluded them for so many years. It wasn't until researchers solved several entirely different questions that progress was made, albeit in a highly circuitous way.

During that same period, scientists had been conducting research aimed at curing pernicious anemia—a deficiency causing a lowered ability of the blood to carry oxygen. This type of anemia is "pernicious" because it was until recently a fatal disease; it had killed Mary Todd Lincoln (the wife of President Abraham Lincoln), Alexander Graham Bell, Annie Oakley, among many others. While a cure was made available in the 1920s in the forms of raw or lightly cooked liver and injections of liver extract, both were very expensive. It was in this backdrop in 1948 that the pharmaceutical company Merck announced it had isolated a red crystalline substance from liver that was thought to be an effective treatment for anemia. Three months later, a team of scientists led by Lester Smith at Glaxo Laboratories in England had isolated the same red crystalline substance independently. Building on the prior naming of vitamins B_2, B_3, B_5, and B_6, the researchers designated it B_{12}. That same year, scientists began testing the new vitamin on patients who suffered from pernicious anemia, and lo and behold, it cured them. Now that the scientists knew what B_{12} was and that it was an effective treatment for pernicious anemia, they needed to find a way to synthesize it cheaply and in large quantities so it could be distributed to the masses. This obstacle was solved when it became apparent that it wasn't the animals but the microorganisms living in the guts of the animals that were the producers of B_{12}.

Drawing upon the recent introduction of penicillin in the 1940s, a compound produced by a mold that was proven to destroy staphylococci (a type of bacteria that causes most staph infections), scientists were eager to identify additional antibiotics that could kill other dangerous organisms. One of these scientists was Benjamin Duggar, who worked at Lederle Laboratories of the American Cyanamid Company, a rival of Merck. According to one of his biographers, J. C. Walker, inspired by others who had found antibiotics produced by fungus in soil, Duggar "began a project which entailed a systematic search for other antibiotic-producing fungi."[11] He and his team procured and analyzed thousands of random scoops of dirt from many of the places he stopped in his travels.

One of these contained a fungus that killed a wide array of harmful bacteria. Duggar named it *Streptomyces aureofaciens*, or "gold-bearing,"[12] because the agent it produced (named aureomycin) was a bright yellow chemical. It was these antibiotic-producing fungi that played the crucial role in mass producing B_{12}. From the analogy that many organisms synthesize B_2 for themselves, researchers suspected other organisms might also be able to produce B_{12}. Some studies have found certain microorganisms isolated from the soil, including several species of *Streptomyces*—one of which was *Streptomyces griseus*—produced vitamin B_{12} as a by-product of brewing streptomycin. The same year, Lederle Laboratories of the American Cyanamid Company announced *Streptomyces aureofaciens* was also found to be a good producer of vitamin B_{12}. Not long after, Merck developed a technique for making large quantities of B_{12} at a low price derived from these microorganisms.

It was a British-American biologist named Thomas Jukes who would finally connect this convoluted chain of discovery of B_{12} to animal feed, yielding an unexpected result that would redefine the meaning of APF. Jukes, a scientist at Lederle with interests and expertise in nutrition and physiology, was eager to evaluate the impact of adding the newly discovered B_{12} to animal feed. So, in December of 1948 he designed a clever experiment. He divided recently hatched chicks into different groups with each one receiving a unique diet. One group received a diet low in nutrients (serving as a control). The other groups received varying levels of doses of different supplements—among them, liver extract, synthetic B_{12}, and "mash," a growth medium containing B_{12} that aureomycin had been brewed in. Upon weighing all the chicks several weeks later, Jukes noted most of the chicks in the control group that received a nutrient-deficient diet died and almost all of the other chicks in the experiential groups not only lived but grew to be 2.5 times heavier than the few in the control that did live, proving the supplements he gave them played a vital role in their underlying health. But he also found something very unexpected: Among the chicks that had lived, those that received the highest amount of mash weighed the most, much more than the chicks that just received the synthetic B_{12}, indicating "vitamin B_{12} did not remedy completely the deficiency of the basal diet."[13] At first, Jukes wasn't sure why this was happening. But after repeated experiments, Jukes confirmed what had caused the chicks to put on so much weight was not the B_{12} in the mash, but instead, a "new growth factor"—trace amounts of aureomycin. Jukes speculated (correctly, as it later turned out) the antibiotic was effective in preventing infections common in animals raised in crowded conditions, and that energy which normally would have been consumed by their immune system was going instead to making larger muscles and bones. All in all, APF was no longer considered to be just B_{12}, but the combination of the vitamin plus a low dose of antibiotic.

The implications of the finding, heralded in a 1950 article by the *New York Times* as "one of the greatest growth-promoting substances so far discovered, producing effects beyond those obtainable with any known vitamin,"[14] were truly groundbreaking. Just five pounds of unpurified antibiotic added to a ton of animal feed increased the growth rate of chickens by nearly 50 percent. Animal feeds laced with the vitamin-drug combination lowered mortality rates, decreased the time needed to reach market weight, produced more meat, and cost less than the animal-based proteins. News spread fast, and only a few years after Jukes's discovery, American farmers were feeding their animals nearly half a million pounds of antibiotics a year. After years of animal feed shortages and costly meat, poultry was on the upswing.

The only step left was to transform the anatomy of the chicken itself, and by the time antibiotics hit the scene, that too was already under way. As World War II came to a close, a man named Howard Pierce had a clever idea about how to bring the cost of chicken even further down while making it look and taste closer to red meat. Pierce was the national poultry director for the country's largest food retailer, the Great Atlantic & Pacific Tea Company (A&P). At what would become a historic 1944 industry meeting in Canada, Pierce shared his vision for "the development of a chicken of a type similar to that of the broad-breasted turkey."[15] At the time, a dressed chicken looked very different from the one we know today; it took a considerable amount of effort merely to extract an edible portion of meat, and it didn't look particularly appetizing. Pierce surmised that if strips of chicken could be made as thick as pieces of beef, consumers might buy more. The growers at the meeting were seduced by his idea, and by the next summer he had the support of the USDA and every major poultry producer in the country. The committee's unifying goal was to breed a better chicken—what they collectively dubbed, the "Chicken of Tomorrow."

At the heart of Pierce's strategy was an XPRIZE-like national contest to engage breeders in identifying the crème de la crème of foul, what the *New York Times* referred to as "a sort of super-bird—more meaty, more juicy, more tender, and less bony than the chicken in last night's pot."[16] First, as part of forty-two regional contests held around the country in 1946 and 1947, contestants submitted fertilized eggs that would hatch under the same conditions, be fed the same kind and amount of food, and be given the same vaccinations. The tastiest and fattest breeds then competed in a 1948 national contest where, after twelve weeks and two days, the chickens were weighed, slaughtered, and dressed as if they were going to be sold. Judges recruited from industry, academia, and government scored the biggest and best birds according to eighteen criteria that resemble those in a human beauty pageant, including body structure and skin color. The owner of the plumpest and fastest-growing chicken received a $5,000 cash prize.

To drum up interest, Pierce launched a slick publicity campaign, including a short documentary narrated by Lowell Thomas, then the nation's most famous newsreel reporter. Aided by visuals of men in white coats and ties who worked in agricultural labs, the message was clear: There was a dire need to modernize the chicken into "a broad-breasted bird with bigger drumsticks, plumper thighs and layers of white meat."[17] To supplement the film, the committee also cosponsored a "Chicken Booster Day," which included a New York City banquet and a screening of *Chicken Every Sunday*, a sentimental comedy-drama featuring famed actress Celeste Holm and a reassuring meal as the real star.

Although a critic at the *New York Times* may have thought the film was "larded with rich and wholesome portions of nourishing Ma-Loves-Pa and . . . seasoned with more than generous sprinklings of standard bucolic farce,"[18] the marketing campaign worked. The regional contest netted forty winning contestants from twenty-five states who sent 31,680 prized eggs to a hatchery in Maryland. After much deliberation, Charles Vantress from California, who submitted a breed made by crossing California Cornish males with New Hampshire females, was announced as the winner. Vantress produced big chickens averaging nearly four pounds—twice the size of a typical chicken of the day—with good feed efficiency, requiring only three pounds of feed for every pound of chicken. The contest was such a success that another one was held three years later, which Vantress won again with another cross breed. Having achieved unprecedented fame in the poultry industry, it wasn't long before most of the commercial broilers raised in the United States were directly descended from Chicken of Tomorrow prize winners. Indeed, by 1951, Vantress's breed encompassed one-third of broilers on the market; by 1959, it was 60 percent.

Inspired to improve upon the winning chicken breeds, the largest vertically integrated poultry companies—often in collaboration with emerging biotechnology firms—further enhanced their genetic makeup.

The optimized results were considered intellectual property, off limits for a period of time to competitors. Through a series of complicated crossbreedings, these companies produced chickens that made Vantress's winning chicken breeds look like scrawny runts. For example, their breasts—which offered more of the white meat consumers desired—were now twice as large. Their muscles grew so fast their bones and tendons were unable to support their own weight. To help them adjust to the conditions of a factory farm, scientists modified their behaviors too. Chickens began to eschew even basic forms of expression like pecking and flying short distances, instead living a shortened life of near catatonic passivity.

In 1941, it took about twelve weeks and 4.2 pounds of feed to produce a pound of live-weight broiler chicken; by 1950, it was 3.3 pounds, and by 1963, only 2.5 pounds of feed were required per pound of gain. Faster gains, further mechanized facilities, and better housing reduced the amount of labor required to produce broilers. In the early 1950s, officials estimated it took 3.1 hours to produce one hundred pounds of broilers; one decade later, it was only one hour. These incremental efficiencies were reflected in the price of chicken over time: in 1950, chicken averaged 57 cents a pound while beef and pork cost 70 and 67 cents a pound, respectively. By 1965, chicken had dropped to about 39 cents per pound, while beef and pork were higher than fifteen years earlier. That same year, chicken consumption reached an average of nearly 25 pounds per person, up another 15 pounds from 1950. Chicken, once shunned from the dinner plate, was now wildly popular. It was also very profitable, and it was only a matter of time before the industrialized production methods pioneered by chicken farmers would characterize all of animal agriculture.

It would take another thirty years before chicken would overtake pork, and forty-five years before it would surpass beef, yet the poultry industry was well on its way. Indeed, another major transformation was on the horizon, one that would make it easier for people to consume all kinds of meat.

6

An Era of Convenience

I don't like to eat snails. I prefer fast food. —*Strange De Jim*

A few years ago, I went to Six Flags Great Adventure with some friends. We arrived early in the morning to beat the lines for our favorite rides. Sometime in between dangling upside down on a roller coaster and getting soaked on a high-speed cruise down a winding water course, we took a break to grab some food. We walked into the first eatery we saw. But as I scanned the menu, I noticed a problem: There wasn't a single vegetarian option. My omnivorous friends kindly agreed to leave and find another restaurant, but we quickly encountered the same problem. After we failed at a third restaurant, my friends started getting hangry. I decided if the next restaurant did not offer a vegetarian option, I would either break my plant-based streak or grab a soft pretzel and bucket of popcorn from a food cart. Mercifully, the next restaurant had a black bean burger, and we all rejoiced.

As we enjoyed our meals and planned the rest of our afternoon, I reflected on our ordeal. Why was meat always the default option? More specifically, how did it become the epitome of dining convenience?

Before the early twentieth century, shopping for meat was a very different experience than it is today. This was in part because shopping for groceries involved trips to multiple specialty shops, including greengrocers for produce, bakeries, fishmongers, dairy shops, dry goods stores, and butchers. A simple dinner could require visits to half a dozen stores and an entire afternoon—that is, until two brothers, George L. Hartford Jr. and John A. Hartford, and the Great Atlantic & Pacific Tea Company (A&P) came along.

The A&P was founded by George F. Gilman, the son of a wealthy leather merchant. After his father died in 1859, Gilman ditched the tanning business and began to sell high-priced tea in bulk from a storefront in New York City under the name of his father's company, Gilman & Company. A couple years later, perhaps in a nod to his ambitions, he renamed it the Great American Tea Company, and by 1869 he had opened eleven stores. In 1870 he created a new business called the Great Atlantic & Pacific Tea Company as part of an initiative to distribute prepacked tea to merchants nationwide via the newly completed transcontinental railroad. It was around this time that George H. Hartford Sr.—who had joined the Great Atlantic & Pacific Tea Company and quickly climbed the ranks—began to shine. Just days after the Great Chicago Fire in 1871, the A&P sent wagonloads of staff and food supplies into the city. Seizing the opportunity, Hartford Sr. directed the purchase of a property that would become the first A&P store outside of New York City, making it one of the first retail chains with a national presence.

By 1875, the A&P had stores in sixteen cities. In 1878, Gilman retired and left control of the business to Hartford Sr., who had proven to be a valuable asset. But Hartford Sr. faced a big problem: By the early 1880s, tea had become such a common staple that its price had fallen dramatically. In response, Hartford Sr. diversified by adding items such as baking powder and sugar to A&P stores; by the 1890s, they were stocked with a variety of canned goods, butter, and soap as the A&P transformed from a tea company to a grocery chain. During this time, his sons George and John began employment at the company, engaging in entry-level tasks while learning the ins and outs of the business from their father. By the early 1900s, they had taken on substantial responsibilities and were shaping the future of the company. In 1912, they opened an experimental grocery store much larger in size that abandoned costly practices such as elaborate furnishings, excess staff, credit, and delivery. These "economy stores" were a huge success. By 1916, one year after the brothers took over the company, A&P's sales had soared from $31 million to $76 million. By 1925 they had nearly fourteen thousand stores and sales of $352 million, more than any retailer in the world.

Soon the brothers embarked on another experiment: They began adding perishable items to A&P stores, including meat. For the first time, it was no longer necessary to make an extra stop at the butcher to purchase meat; now it was available at a counter-service meat department alongside other grocery staples in what became known as combination stores. That same year, the A&P became the first retailer to have revenues of $1 billion in a single year, 10 percent of which came from meat sales.

By 1936, the brothers had begun to invest in a new type of store that made combination stores of the time look miniature: supermarkets. The supermarket was not their idea but that of a man named Michael Cullen. Cullen was born in Newark, New Jersey, in 1884 as the child of Irish immigrants. At the age of eighteen, he joined A&P, where he rose from clerk to divisional superintendent. After an impressive seventeen years at the company, he left A&P to become general sales manager for Mutual Grocery and later the Bracey-Swift stores in Illinois and Missouri. But it was while working at Kroger that he conceptualized the supermarket. By driving slightly out of the way and tolerating a no-frills warehouse-like environment, Cullen reasoned consumers could enjoy lower prices and a wider selection of offerings. In 1929, he proposed the idea to the company's president in a letter, arguing so many consumers "would break . . . [the] front doors down to get in."[1] Nevertheless, his bold ideas were ignored. Yet Cullen was so convinced it would work he quit his job at Kroger and moved his family to Long Island to launch the "monstrous" store of his dreams. On August 4, 1930, King Kullen Grocery Company store—the first true supermarket—opened its doors, and it was an overnight success.

Hot on the heels of King Kullen, A&P opened eleven hundred supermarkets within a matter of years, in the process closing many of its smaller, specialized stores. Sales at the average A&P store more than doubled, and the amount of merchandise moving through the average store nearly quadrupled. It was during this time that the Hartford brothers doubled down on A&P's investment in meat by opening the first "self-service" meat departments in select stores. Here's how self-service meat departments worked: In each store, butchers cut, weighed, and packaged cuts of meats in a back room in preparation for the day's sale and displayed them in self-service cases. A clerk supervised the products and helped consumers who were unfamiliar with the process of buying self-serve meat. The new operation increased meat sales in the experimental combination stores by about 30 percent. Building on the success of the pilot, the brothers gradually rolled out the new self-service meat system throughout the early 1940s.

Although the integration of meat and the grocery store—first in the form of a full-service counter and then in the form of self-service—radically transformed the experience of buying meat, it wasn't without challenges. Particularly in these new self-service meat environments, quality control was a huge challenge. Despite refrigeration technology having significantly improved since the heyday of Swift and Armour thanks to the introduction of modern refrigeration, commercial refrigerators were still unreliable. When temperatures rise even just a few degrees above 35°F, microorganisms begin growing and multiplying and enzymes break down the tissue, causing uninviting discoloration. Cellophane—which was first invented by Swiss chemist Jacques Edwin Brandenberger and later refined by chemical manufacturing company DuPont in the United States during the early 1900s—was also in its infancy, as it often became brittle and nondurable at low temperatures, preventing fresh meat from receiving the precise amount of oxygen and moisture required to attain the red color and juiciness consumers desired. These issues prompted one grocery store owner to note in the mid-1940s that until "properly refrigerated transportation and display equipment is available, peak 'farm-to-table' freshness cannot be maintained."[2] For this reason, self-service meat departments were the exception and not the rule; in 1948, only 39 percent of independent grocery stores and 56 percent of chain stores offered them.

Improvements in both commercial refrigeration and in cellophane were spurred by World War II, during which scientists were tasked with finding innovations to feed soldiers as inexpensively as possible. This meant improving cold storage warehouse technology to ensure minimal loss and waste due to unnecessary spoilage. In collaboration with private industry, cold storage significantly improved, and the investment paid off: Only 0.1 percent of food purchased by the War Food Administration (WFA) to feed the troops went bad. Although retail refrigeration lagged behind, many of these underlying technical enhancements eventually were applied to this new link in the cold storage chain. A similar story unfolded for cellophane, which was designated an "essential material" in part because of its use in packaging rations for soldiers, fueling additional research on how to improve it. By 1946 DuPont had successfully created an improved version of cellophane, striking an ideal balance of moisture control and oxygen penetration—one sufficient for, among other uses, packing meat in self-service refrigerators.

The impact of these innovations on self-service meat was dramatic: In 1946, there were only twenty-eight supermarkets with complete self-service meat departments; by 1953, more than 50 percent of all supermarkets had them. As improvements in commercial refrigeration and packaging continued throughout the second half of the twentieth century, self-service meat became the overwhelming norm.

During this same period, domestic refrigeration also improved dramatically, making it possible for consumers to store meat for longer periods of time without worrying about spoilage. The earliest household refrigerators were known as "iceboxes," and they ranged from plain wooden boxes to intricately carved pieces of furniture. Here is how they worked: A large block of natural or artificial ice purchased to size was placed in a container at the top of the device. It gradually melted, and water trickled into a tray at the bottom. The ice gradually cooled the air inside and it circulated down to a larger storage chamber.

In the 1920s, the first true mechanical refrigerators became available, but they used combinations of toxic liquids and gases as refrigerants, such as ammonia, methyl chloride, and sulfur dioxide. Electric refrigerators were introduced in the 1920s, but like any new technology, they were expensive and mainly used by the wealthy. Yet prices swiftly dropped: At the start of the 1930s, just 8 percent of American households owned mechanical refrigerators; by the end of the decade, the average price of a refrigerator had dropped by half, and 44 percent of consumers owned one. According to Jonathan Rees, by the 1940s, all makes and price ranges of refrigerators were improving; they were quieter and lighter, required less energy, leaked less frequently, had achieved adequate temperature control, and even had a light that went on when opened. By the 1950s, nearly every home in the United States owned a refrigerator.

Another innovation that made meat consumption more convenient involved stripping meat off the bone and processing it. As Anastacia Marx De Salcedo, author of *Combat-Ready Kitchen, How the U.S. Military Shapes the Way You Eat*, explained to me, up until World War I, beef was shipped dressed, much like it was by Swift and Armour half a century ago: "The military was shipping over these carcasses of cattle and so forth and it took a long time . . . so they wondered if they could instead of shipping over entire carcasses, take the meat off the bone and pack it up into a box."[3] The idea was that without its bones, fat, and cartilage, it would be easier to transport and take up less space on crowded trains and ships. In 1918 the Army set up a "boxed-beef" processing plant and distribution center in Chicago, where meat was broken down into basic deboned cuts such as rounds and loins before being shipped in containers; any remaining bits were cubed for soups and stews or ground into chopped meat. But without proper deboning techniques, much of the meat remained on the carcass, and the meat that didn't suffered an even worse fate than what could be found in the refrigerators of the early days of combination stores and supermarkets: "The results were not that good . . . for those very first soldiers eating meat that had been cut off the bone in camps in Europe," Marx De Salcedo told me.

The Army continued its research on deboned beef throughout the 1920s and 1930s, albeit on a very small scale, in part due to the lobbying efforts of Dr. Jesse H. White, who had supervised the production of deboned beef during World War I for the Navy. But it wasn't until 1938 that he had a breakthrough, when the Army successfully enlisted the help of none other than Swift and Armour. With their assistance, by the end of the decade, the Army had successfully developed a deboning technique far superior to dressed beef. Soon after, the government began to sponsor research on deboning other types of meat, including lamb and pork.

Simultaneously, private meatpackers began using this emerging technology to create a wide variety of new convenience foods. Consider the origin story of Spam, perhaps the most iconic brand in the category. Spam was the brainchild of Jay Hormel, an army lieutenant who helped experiment with deboned beef in World War I. His father, George Hormel, had founded Hormel Foods Corporation in 1891. George Hormel's greatest expansion came just two years later during an unlikely time: the Panic of 1893—an economic depression that lasted until 1897. Hormel reasoned financially stretched consumers might be willing to eat cheaper cured and smoked meats. He was right: When the company released new inexpensive products like thinly sliced Canadian bacon, business boomed.

When Jay Hormel became president of the company in 1929 during the Great Depression, he faced an even worse economy. Like his father, Jay Hormel too figured consumers would embrace highly processed meat products if they were substantially cheaper. Like his father, he too was right. Though processed foods kept the company afloat, the prolonged Depression was cutting deeper and deeper into sales. Soon, Hormel had another idea. For decades, the company had discarded thousands of pounds of pork shoulder because they required a lot of effort and specialized skill to debone and because consumers preferred meat that looked like what was found at their local butcher shop. But what if he could transform the loose pork shoulder and fat trimmings into something more familiar, like a meat loaf?

Under his direction, Jay Hormel's food scientists went to work with the aim of developing a loaf in the shape of a rectangle block—one that disguised its true origins and was large enough to feed an entire family with leftovers for sandwiches the next day. Finally, after much trial and error, they perfected both the new process and the specially designed machine. On May 11, 1937, after conducting a naming contest, the company registered the trademark for "Spam"—a play on words for spice and ham (despite neither appearing in it). Spam was made of five ingredients: chopped pork shoulder meat "with ham meat added,"[4] salt, water, sugar, and sodium nitrate (the kind of salts used to cure meats like bacon and hot dogs). By 1940, 70 percent of urban homes in the United States had used Spam.

But it was World War II that made this "miracle meat in a can" a massive success. The military needed to transport meat to frontline troops without it spoiling, and Spam met that need in spectacular fashion: It was cheap, portable, and didn't need to be refrigerated. For these reasons, the military became the largest purchaser of Spam, responsible for doubling its sales between 1939 and 1942. By the end of the war, over 150 million pounds of Spam was consumed by the armed forces.

It was also during World War II that another category of convenience foods became more popular: frozen foods, encompassing everything from various cuts of meat (such as steak, haddock filet, and oysters) to fruits and veggies (such as raspberries, peas, and spinach). The idea was simple: Rather than spend the time required to cook it, why not pass that labor onto someone else and merely defrost it on the stove or in the oven? The timing was no coincidence: With nearly all tin and aluminum production diverted to the war effort, there was less canned food available, which led some people to try frozen food for the first time. Among those who did were women who had taken the factory jobs; with less time for domestic responsibilities, they were looking for a quicker and easier way to prepare meals. In 1930, only eighty thousand pounds of Birds Eye Frosted Foods products were sold; by the mid-1940s, sales had surged 1,000 percent. All told, Americans bought eight hundred million pounds of their food in frozen form between 1945 and 1946.

However, demand for frozen foods waned in the years after the war. At first, this was because supermarkets avoided investing in expensive freezer displays, mainly because frozen meals were expensive and advertised as luxury foods to consumers. But even when supermarkets agreed a freezer section was essential and the price of frozen foods fell, they still lagged far behind their fresh counterparts. The reason was not complicated: Few Americans owned a freezer. Consider that in 1952, there were thirty-three million household refrigerators in operation, but only four million freezers. Unlike today, early fridges had freezer compartments that barely had space for a few ice cubes, let alone room for entire frozen meals. Additionally, with men returning to their factory jobs after the war, women were largely sent home and no longer had to rely on quick and easy meals to feed their families.

In the early 1950s, food companies began investing their energies in a new product they hoped might entice consumers to jump on the frozen food bandwagon: complete frozen meals. The idea was not new: An engineer named William L. Maxson is credited with inventing them during World War II for military passenger planes that routinely crisscrossed the Atlantic Ocean. With these slow flights taking the better part of a day, in-flight food service was a necessity. Initially, the crew and passengers subsisted on cold sandwiches prepared in-flight, but they quickly found frozen precooked meals were much easier to manage as they only required heating. Maxson's company, the W. L. Maxson Corporation, marketed them as "Strato Meals," which consisted of a hunk of meat and two vegetables arranged on a tri-partitioned, paper-fiber container. Within two years, five hundred thousand meals had been served on these planes, a statistic that prompted Andrew Hamilton of the magazine *Popular Mechanics* to make a bold prediction: "Before long you may see frozen dinners served in hotels, trains, planes, ships, factories, offices and your own home. They will probably be sold in grocery stores and delicatessens."[5]

It was one company's miscalculation and an employee's clever idea that ultimately altered Americans' opinions about convenience foods and gave rise to a prosperous frozen meal industry. The origins of this new industry begin with a culinary folklore involving five hundred twenty thousand pounds of turkey. In 1953, C.A. Swanson & Sons, a frozen-food company based in Omaha, Nebraska, had vastly overestimated turkey sales for that year's Thanksgiving. This left the company in a predicament: either chuck the birds and lose a lot of money or find a way to sell them in one form or another. With nowhere else to keep the overflow and with no immediate ideas on how to sell them, C.A. Swanson & Sons loaded the turkeys onto ten train cars equipped with freezers. In order to keep the electricity flowing, the train had to move continuously, so it aimlessly traveled back and forth between the Midwest and the East Coast.

At its wit's end, the company desperately asked its employees to submit ideas for solving the turkey problem. According to legend, the winner was a salesman named Gerry Thomas, who proposed to Swanson they put the turkey meat into frozen dinners. The company then got creative by packing the turkey into colorful boxes made from laminated parchment designed to look like wood-grained television sets, complete with simulated tuning and volume control knobs. They called their product the "TV Dinner." In 1954, Swanson launched the TV Dinner nationwide. At a price of 98 cents per unit, the company sold over ten million TV dinners by the end of the year. Swanson soon added new meals to its lineup including fried chicken, roast beef, and filet of haddock. By 1959, Americans were spending half a billion dollars a year on frozen meals and some two hundred fifty million TV dinners had been consumed.

When microwaves became a fixture of kitchens in the 1980s, they helped catapult the consumption of TV dinners. Like the many innovations preceding it, the microwave oven was invented by accident. Percy Lebaron Spencer was a self-taught engineer who worked at Raytheon, a contractor for the US Department of Defense. In 1939 he was experimenting with magnetrons—vacuum tubes that produce microwave radiation used in radar systems—when he noticed a peanut cluster bar in his pocket had melted. He decided to see if something similar happened with popcorn; it didn't melt, of course, but it did pop. In another experiment, he placed an egg in a teakettle with the magnetron placed directly above it. One of his doubting coworkers made a foolish decision to open the kettle to observe the egg, and was rewarded with literal egg on his face. Spencer next built a metal box and fed microwaves into it, discovering in the process that they heated foods much more rapidly than a conventional stove or oven. On October 8, 1945, Raytheon filed a patent for the first microwave oven, eventually naming it the "Radarange."

Two years later, the world saw the first commercially produced microwave oven. It was shorter than the typical basketball player (though not by much), weighed about three quarters of a ton, and cost $5,000. Though consumers were skeptical at first, by 1986 a quarter of American households owned these "miracle workers." Even in 1979, the technology had progressed enough that Swanson announced a new line of frozen meals that were reformulated and repackaged—using plastic trays instead of aluminum—for microwaving. Now, with the touch of a couple of buttons, the TV dinner could be defrosted in just a few minutes. By 1983, sales of frozen dinners reached $750 million and sales of frozen entrees (which consisted of a starch instead of a green veggie alongside the meat) reached $1.9 billion. Commenting on these figures, one industry executive at the Campbell Soup Company, which at the time owned Swanson, remarked in 1984 to a *New York Times* reporter, "Frozen foods used to be millstones around the big guys' necks. . . . But now they're hurling trees and mountains at each other to take over."[6] Frozen meal sales grew practically every year between 1953 and 2008. Though the industry peaked in 2008, Americans still spent more than $10 billion on frozen dinners in 2020, and today more than 99 percent of households have frozen meals—many with meat as the main—in their freezer.

But well before frozen meals and microwaves became all the rage, an entirely new trend of convenience had started to take over: fast-food restaurants. In the nineteenth century, the closest thing resembling a modern-day fast-food restaurant was the cafeteria. Cafeterias emerged as part of an adaptation to a new, fast-paced lifestyle in cities like New York, Chicago, Los Angeles, and San Francisco. The value proposition was simple: Workingmen could eat a relatively inexpensive lunch in a hurry. Embodying the same chaotic energy that filled the city streets, cafeterias were typically jam-packed with hungry customers, most of whom ate quickly and often while standing. And much like modern fast-food restaurants, the intellectual elite criticized cafeterias for putting efficiency above manners and quality.

In 1902, Philadelphia-based entrepreneurs Joseph V. Horn and Frank Hardart opened the first US automat, a cafeteria with some added bells and whistles. Also known as "vending machine restaurants," automats offered prepared foods behind small glass windows and coin-operated slots. Here is how it worked: Once customers made their selection from the window display, they simply inserted the required number of coins, turned a knob, opened the door, and removed the food. The automation allowed the duo's cafeterias to operate with the same efficiency as a conventional cafeteria but without any congested lines.

In addition to instilling in customers a new expectation for efficiency, automats set a new basis for standardization and predictability. In most cafeterias, standardized food simply meant equal portions and prices, but Horn and Hardart also rigorously controlled the recipes, preparation process, and presentation through regular sampling and detailed training manuals. They even regulated how big the bacon on top of the beans should be, exactly how long to cook it, when to turn it over, and how to position it on a plate.

Horn and Hardart's Philadelphia location was moderately successful, but it paled in comparison to the sensation that was the automat they opened at Broadway and East 14th Street in Manhattan on July 7, 1912. The cafeteria benefited from the city's unique subway system, which provided a constant stream of pedestrians shuffling past the automat windows. Automats truly thrived in the commercial districts of Manhattan, especially during the Great Depression, when they offered a tasty meal at a reasonable price. A single Horn and Hardart outlet served on average ten thousand customers per day. But as an era of prosperity began after World War II, automats—known for being a relatively low-end option—lost their appeal as customers flocked to the suburbs. The automobiles that made suburban life possible were simply incompatible with automats. (Horn and Hardart eventually replaced their dying restaurants with Burger King franchises, proving that the fast-food world is a small one.) Within the decades to come, an emerging trend would lure away the remaining automat customers and put the fast-food industry into overdrive.

The first modern fast-food outlet can be traced back to a man in Wichita, Kansas, named J. Walter Anderson. Anderson was born to Swedish immigrants on a farm outside the town of St. Mary's, Kansas, in 1880. As a young adult, Anderson embraced a care-free attitude. He bounced from one college to another before dropping out for good, at one point working as a janitor and living in an abandoned house. He spent years wandering throughout the Midwest, taking on odd, low-commitment jobs like washing dishes and cooking in restaurants. Worried about his son's lifestyle, Anderson's father bought him his very own restaurant in Marquette, Kansas, in 1905. But within a year Anderson had sold the restaurant so he could pursue his dream of organizing a traveling stage show; it failed within weeks.

Over the next decade Anderson took another series of a low-paying jobs. While employed as a short-order cook in a local diner, Anderson began experimenting with grilling and preparing ground meat in novel ways. His customers raved about his flattened patty that was seared on both sides and served on a bun instead of bread slices. Legend has it Anderson stumbled upon the idea when he angrily flattened a slow-cooking meatball with a spatula and realized it cooked faster this way, and thus the modern hamburger was born. Inspired by his creation's popularity, Anderson decided to give the restaurant business another go. After multiple failed attempts, he received a small loan to buy and repair an old shoe stand. He installed a small counter and three stools, a griddle, and various cooking utensils. Over the door he hung a sign reading, "Hamburgers 5¢." With no money left to buy food supplies, Anderson convinced a local grocer to loan him enough beef and buns for the first day of the operation. It was a massive success, the hamburgers selling out within hours. His hamburger stand continued to prosper, and over the next four years he added three more locations in other Wichita neighborhoods. What exactly was the appeal? As food historian Elizabeth Rozin eloquently writes, "The meaning of the burger is as a kind of common denominator of the beef experience, with all the flavor, aroma, tenderness, and juiciness in a cheap and accessible form. The meatiness, the beefiness, the succulence of the fat are all there in that unassuming patty. For perhaps the first time ever, the hunger for all that beef represents, [could] easily be satisfied, available to almost anyone."[7]

Despite his success, Anderson's interest in further expanding his hamburger haven met with obstacles. There was still a persistent stigma associated with ground meat, largely due to Upton Sinclair's exposé. Meanwhile, affluent customers tended to view hamburgers—especially those cooked in poorly constructed buildings on busy street corners like Anderson's—with scorn, relegating them to the working-class diet. In the early years of the Prohibition era, some even suspected food stands were fronts for speakeasies or brothels.

When Anderson tried to lease an additional property from a local dentist, the man insisted Anderson provide guarantees. The insurance and real estate broker who was negotiating the deal for Anderson, Edgar Waldo Ingram, became intrigued with his operation. In fact, he was so impressed with his customer base and growth potential for the business that he agreed to cosign the lease on the dentist's property and invest his own funds. Shortly after, Ingram sold his stake in his insurance and real estate firm so he could focus full-time on building the country's first modern fast-food outlet.

Anderson knew how to make hamburgers his customers liked, but his business approach wasn't all that different from other food stands and diners in the city. Helping Anderson's business stand out among the crowd of the competitors was where Ingram shined. To boost the public image of their burgers, Ingram created a new name for the company, combining two words that conveyed purity and strength: White Castle. This was a reference to one of the few landmark buildings that survived the great Chicago fire of 1871—a building that served as a model for White Castle's white-washed turret-and-tower design.

Ingram also standardized the food preparation, appearance, and operation of their fast-food restaurants. Each new restaurant had the same layout: a grill, a counter, five stools, and two male employees. The menu featured a streamlined selection of hamburgers, coffee, soda, and pie. To ensure the food arrived quickly and the same every time, Ingram implemented a system wherein cooks churned out identical small, square burgers on a large scale. First, ground beef balls were placed on the grill and topped with shredded onions, then they were flipped and squished into a patty before buns and pickles were added. Though a precursor to the fast-food assembly line, this method helped to usher in the fast-food industry as we know it today.

To convince the public that White Castle's hamburgers were made quickly and were safe to eat, Ingram leaned heavily on every public relations channel at his disposal. He insisted utensils be kept squeaky clean, and claimed to use only special cuts of shoulder meat (with no trimmings from any other less desirable part). To underscore White Castle's guarantee of freshness, Ingram had meat delivered from a local butcher several times per day while promising that leftover meat was never used. Ingram added to the sense of transparency by placing the meat grinder in full view of customers in the dining area. Employees needed to be well groomed with clean uniforms. Years later, Ingram would go even further by establishing a test kitchen and quality control laboratory where "scientific" studies were conducted to "prove" the healthfulness of ground beef.

Ingram's efforts to establish White Castle as a high-quality, clean, and efficient eatery were rewarded. Within a year there were many profitable locations throughout the city, and in 1923 they added stores in Kansas and Nebraska. As more outlets opened in other cities, sales naturally increased. Before the end of the 1920s, there were over forty White Castles in the Midwest, as well as several in the Mid-Atlantic region. There are about four hundred White Castles now, which is significant (and also appropriate, given its link to the stoner comedy *Harold and Kumar Go to White Castle*), but pales in comparison to other major fast-food chains. Burger King, for example, has more than seventeen thousand restaurants, and Subway has nearly forty-five thousand. To understand one of the major innovations that constituted this difference in scale, we must meet a man named Roy W. Allen.

There isn't much information on Allen's early life other than that he was an Illinois native who headed west to buy, refurbish, and sell old hotels. Along the way he met a chemist who claimed to have perfected a recipe for improving the taste of root beer. Allen market-tested the root beer at one of his hotels and was impressed by how much customers liked it. He bought the rights to manufacture and market the new root beer syrup—a secret blend of herbs, spices, barks, and berries—and opened a walk-up root beer stand in Lodi, California, in 1919. Three years later, Allen made Frank Wright, one of his employees, a partner in the business, and they aptly named it A&W Root Beer.

In 1923, Allen introduced what some historians believe was the first drive-in restaurant in Sacramento, California, featuring "tray boys and tray girls" (later called "carhops") for curbside service. One year later, after opening more A&W stands in Utah and Texas, Allen purchased Frank Wright's stake in the business.

Then, in 1925, Allen did something groundbreaking: He sold others the right to make and sell his root beer and prominently display the chain's "bull's-eye and arrow" logo and branding. Rather than set up an unknown business from scratch, people could rely on A&W's sound reputation and proven track record. Whereas White Castle owned all of its business locations, A&W had independent owners operate individual stores. In other words, Allen was the first to successfully franchise a food operation.

The result was unprecedented growth. By the mid-1930s there were nearly two hundred A&W outlets. At their peak in 1974, there were twenty-four hundred. But unlike modern franchises, which involve strict standards so that the menu looks and tastes the same no matter where you are in the country, A&W's franchisees had near complete autonomy. Some added other food items to the menu, including popcorn, pork tenderloin sandwiches, hot dogs, and hamburgers.* They opened locations wherever they wanted in any kind of building, and painted them whatever color they fancied. While the public came to expect uniformity in outlets, A&W stubbornly stuck to its business model and dropped to fewer than five hundred restaurants by the mid-1980s. In its place, a mega-fast-food company would take hold.

When we think of fast food, it is McDonald's that springs to mind. By improving on innovations developed by its predecessors and adding some

---

* A&W estimates there are roughly thirty-five thousand menu variations across their restaurants.

new ones of its own, it transformed the fast-food industry into the titan it is today. But first, some history: In 1930, not long out of high school, brothers Richard and Maurice McDonald—names you might know from the film *The Founder* starring Michael Keaton—moved from snowy Manchester, New Hampshire, to sunny Hollywood, California. They were in search of a new opportunity, eager to avoid the same fate as their father, who had lost his job as a foreman in a shoe factory during the Great Depression. They first found manual labor gigs at movie studios, including the Columbia Film Studios back lot, mostly as set movers and handymen. Intrigued with the burgeoning industry, they spent their savings on opening a movie theater in Glendale. It never saw a profit over its four-year run. But the brothers continued to search for a promising entry-level business opportunity, and they finally found it in the drive-in restaurant.

The year was 1937, and the automobile was becoming a fixture in California. Those who owned a passenger vehicle developed an extraordinary dependence on it. Many entrepreneurs capitalized on the trend by building restaurants that catered to curbside service, in which customers parked on the street and a runner took his or her order and then brought food to the car.

The McDonald brothers wanted in on the action. In 1937 they opened "The Airdrome," a small food stand in Pasadena that primarily sold hot dogs via curbside service. Three years later, the brothers transformed the food stand into a larger, but far from roomy, drive-in restaurant in San Bernardino named "McDonald's Bar-B-Que." The seating was all outdoors and consisted of counters and a few stools. Following the yet-to-be-coined mantra "sex sells," their female employees dressed in majorette boots and short skirts. As carhops, they took orders and brought food to the customers who pulled into the parking lot. The menu was almost unrecognizable from what you'd find in a McDonald's today. It had twenty-five items, including tamales, chili, and peanut butter and jelly sandwiches. The restaurant became an overnight sensation, netting the brothers a small fortune.

By 1948, despite their success, the McDonalds grew dissatisfied with their restaurant. Some days they were bored with not enough to do; other days, they were inundated with tasks they didn't enjoy overseeing. They had to constantly hire new carhops and short-order cooks, and a large percentage of their customers were teenagers who broke and stole dishes and glassware. To make matters worse, imitators were popping up seemingly every day. So, the brothers decided to reinvent their restaurant before it was too late.

They closed their doors for three months and implemented changes that completely overhauled their business. They fired the employees who took orders at windows and brought food to cars, delegating some of that labor to customers themselves who would now have to walk inside to place their order and get their food. To avoid the need for a paid dishwasher, and to stem financial losses from the theft of valuable silverware, the brothers got rid of plates and metal cutlery in favor of paper wrapping and paper cups. Their menu became much less eclectic, streamlined to nine items in order to focus on their primary moneymaker: burgers. Hamburgers, cheeseburgers, three soft drink flavors in twelve-ounce doses, milk, coffee, potato chips, and pie were the only items to choose from. (A year later, they stopped selling potato chips and pie and began selling fries and milkshakes.) The brothers initiated a radically new method of preparing food that resembled a Henry Ford assembly line. Each worker performed a different task: One person grilled hamburgers, another person dressed and wrapped them, and so on. The brothers called it the "Speedee Service System." The restaurant—now named simply "McDonald's"—opened on December 12, 1948. Within six months, this new concept driven by extreme efficiency revolutionized their restaurant business, effectively lowering prices and raising the volume of sales. Just two years after reopening, the brothers saw their already healthy profits double.

In 1952, they decided to experiment with a franchise model. Neil Fox, an independent gasoline realtor, became their first franchisee. The brothers hired an architect to design a new building to serve as a prototype for the chain they now wanted to build. They had two goals: create a more eye-catching appearance, and further improve efficiency. To help make it easy to spot their restaurant from the road, they insisted the architect design huge arches that towered over the roof—an idea he so loathed that he quit the project. The brothers proceeded anyway, and once the building was constructed, they added a neon sign on the roof in the shape of the letter "M." Inspired by ahead-of-their-time behavioral economics, the brothers turned off the heating to prevent people from staying too long;

installed fixed and angled seating so customers would hunch over their food and eat faster; and spread the seats far apart to dissuade groups from dining together. People from all over the country went to visit the marvel that was McDonald's. The brothers were getting so many franchising requests they never had to look for candidates. Yet by 1954, the conservative duo only had ten McDonald's franchisees.

That year, the McDonald brothers got the attention of Ray Kroc, then a salesman of Prince Castle–brand Multimixer milkshake machines, when he learned their original San Bernardino location was busy enough to need eight of his machines. When he visited the McDonald's, he was blown away: "I felt like some latter-day Newton who'd just had an Idaho potato caromed off his skull," he later wrote in his autobiography *Grinding It Out: The Making of McDonald's*. "I was just carried away by the thought of McDonald's drive-ins proliferating like rabbits."[8] The McDonald brothers were less ambitious. But Kroc, with an intuitive knack for salesmanship, was persistent, and ultimately convinced them to sell him the right to franchise McDonald's nationwide. The agreement gave Kroc a hefty 1.9 percent of the gross sales from franchisees, and of that the brothers would get 0.5 percent, leaving Kroc with 1.4 percent. The catch was that Kroc had to follow the brothers' plans down to the last detail.

Kroc went to work. Demonstrating his belief in McDonald's, he sold the first franchise to himself under the "McDonald's System, Inc." on April 15, 1955, in Des Plaines, Illinois, near Chicago. This was a test case, as well as a model for future McDonald's franchises. Unlike other franchised brands, Kroc devised a system that put the franchisees financial position ahead of the parent company. Traditionally, franchisers profited on the sales of supplies to the franchisees. But this had an inherent problem: The franchiser made most of the money before the restaurant even opened, and the quality of the restaurant often suffered as a result. Instead, Kroc concentrated on increasing total revenues of all his franchised restaurants—a win for McDonald's and for the franchise. Once the Des Plaines restaurant had proved this new business model was profitable, Kroc went hunting for more McDonald's franchisees. Relying largely on his charisma and word of mouth, Kroc grew McDonald's to over thirty restaurants by 1958. The same year it sold its one hundred millionth hamburger. One year later, he opened more than twice as many new restaurants, bringing the total to over one hundred locations. And the money poured in with them.

However, the relationship between Kroc and the McDonald brothers quickly soured. The McDonald brothers were attached to how they had run their own restaurant, and thus hesitant about messing with success. They demanded Kroc stop making any significant changes to their business model without express written approval from them. Kroc was not too fond of this requirement, especially since the brothers were so reluctant to allow any changes. "It was almost as though they were hoping I would fail," Kroc later wrote. So, he ignored them. With tensions mounting and both sides threatening legal action, in 1961 the McDonald brothers at last decided to sell Kroc the company outright for $2.7 million, a sum Kroc borrowed from investors. It was risky, but Kroc would soon discover it would pay off in spades.

Under the unimpeded helm of Kroc, McDonald's growth exploded. Unlike White Castle restaurants, which were often derisively called "truck stops" despite being largely located in urban centers, Kroc aimed for a family crowd. Leaning on his penchant for promotion, he overhauled McDonald's mascot, a man named "Speedee" with a chef's hat and a sly wink plastered on a hamburger-shaped head. "Ronald McDonald," the oddly charming clown, replaced Speedee in 1963. In television commercials that appeared in 1971, Ronald McDonald inhabited a fantasy world called McDonaldland and had adventures with his colorful friends including Mayor McCheese, Hamburglar, and the Fry Kids. The logic was pretty straightforward: "A child who loves our TV commercials and brings her grandparents to a McDonald's gives us two more customers," Kroc wrote. "This is a direct benefit generated by advertising dollars."[9] In 1976, they produced a line of six-inch-tall action figures to celebrate the iconic McDonaldland characters—a precursor to the countless branded toys included in Happy Meals shortly after their debut in 1979. The Happy Meal concept was pioneered by an ad executive named Bob Bernstein, who had noticed that when parents brought their children to McDonald's, they all tended to share the same meal. Bernstein suspected kids would insist on having their own special meal if one were available. Reflecting on his own son who read the cereal box every morning, Bernstein said in an interview that he suggested to McDonald's that they put "at least ten things on the bag that a kid could read."[10]

In addition to the company's skillful marketing, McDonald's excelled at responding to consumer demand. In the early 1960s, a local franchisee in Cincinnati realized he was losing business to a competitor who sold halibut. The restaurants were in a predominately Roman Catholic neighborhood, and many potential customers were electing to eat fish instead of red meat on Fridays. After a successful test in the franchisee's restaurant, the Filet-O-Fish sandwich made its nationwide debut in 1962, advertised as "the fish that catches people."[11] Three years after McDonald's went public, in 1968, the Big Mac made its debut, and within a year accounted for nearly 20 percent of all McDonald's sales.

McDonald's spectacular growth continued in the early 1970s, reaching over fifteen hundred locations across every state. At no point in its entire history, up until then, had McDonald's failed to make enormous profits. Yet McDonald's would finally hit a snag. The mid-1970s brought a recession, abruptly ending the post–World War II economic expansion. As Americans battled high unemployment, high inflation, and surging gasoline prices, the McDonald's bottom line began to suffer.

A man named David Wallerstein had the answer. As a young executive in the theater business, Wallerstein had once faced the same problem. At the Balaban Theaters chain in the early 1960s, Wallerstein had tried just about every trick in the book to drive sales of high-markup snacks like popcorn and soda with two-for-one specials, combo deals, matinee specials, and so forth. None of these promotions seemed to work. Then, one evening, Wallerstein had a realization: People simply did not want to buy two or three popcorns; carrying that much food was inconvenient and looked gluttonous. What if, instead, the portion sizes of popcorn were much bigger and the price only a little bit more? He created jumbo-sized containers and put up signs to advertise them. The results after the first week were fantastic: Individual sales of popcorn rose sharply. Drawing upon this experience, Wallerstein suggested to Kroc they sell a jumbo-size bag of fries. But Kroc was skeptical: "If people want more fries," he said, "they can buy two bags."[12] To convince his boss, Wallerstein conducted his own survey of consumer behavior. He visited various McDonald's in Chicago and researched hundreds of customers. He noticed they often ate the entire bag and then scraped and pitched around the bottom of it, desperate for more salty goodness. When Wallerstein shared what he saw with Kroc, he finally gave in. Within months of introducing jumbo sizing, more customers were visiting McDonald's, and each one was spending more.

What started with fries was soon applied to other menu items, as it was impossible to put the super-sized genie back in the bottle. By the mid-nineties, one out of four meals sold at fast-food restaurants was a value meal, a special combo, or a supersize meal. In 1955, a McDonald's hamburger patty weighed 3.7 ounces. Today one weighs approximately 7.3 ounces, a Big Mac 7.5 ounces, and a Quarter Pounder and Quarter Pounder Deluxe a whopping 9.2 ounces. According to the Centers for Disease Control, a typical fast-food burger weighed just 3.9 ounces in the 1950s. Now, it's more than three times larger. It may be no coincidence that our daily calorie intake has expanded alongside the introduction of supersize portions. The total caloric intake of the average American grew from 2,109 calories in 1970 to 2,568 calories in 2010. As Pew Research put it, that's "the equivalent of an extra steak sandwich every day."[13]

Though McDonald's enjoyed even greater success after the recession ended in 1975, the fast-food chain was about to encounter yet another problem. Throughout the 1950s consumers had become increasingly concerned about cholesterol, responding to the scientific community's encouraging consumers to cut back on red meat. That concern ratcheted up in 1977 when the US government offered radical new dietary advice. For the first time, the government urged its citizens to "decrease consumption of meat and increase consumption of poultry and fish."[14] Executives at McDonald's—a hamburger restaurant at its core—were understandably worried. Their attempts at adding chicken items to their menu before had all failed as badly as Ray Kroc's infamous "hula burger," a meatless burger consisting of pineapple and cheese on a bun, but now it was imperative they find a winner in poultry.

McDonald's management leaned on Rene Arend, its executive chef in charge of corporate product development efforts, to come up with the answer. According to legend, McDonald's president Fred Turner and Arend passed each other in the hallway when Turner casually suggested that he experiment with making a chicken nugget. That same day, Arend cut up some chicken into little bits, covered it in a white flour batter, and dunked it in hot fat. After pulling it out and letting it cool off, he took a bite and was instantly in love. The nuggets—small pieces of breaded, deep-fried reconstituted chicken held together by stabilizers—were hardly a health food, but technically they were poultry. In 1979, McDonald's began testing the Chicken McNugget in select markets, which soon proved a massive success.

In order to introduce the wider nation to the McNugget, McDonald's needed a massive supply of chickens. No company had ever attempted to mass-produce chicken at this scale before. The solution came from a man who had long been ignored in his attempts to do just that. In 1967, Donald Tyson took the helm of his father's eponymous company with a singular aspiration: to make it one of the world's leading food producers. The future of Tyson Foods had just one problem: Chicken wasn't on the menu of most fast-food restaurants. Though Tyson had broken ground by selling boneless breasts (drawing upon similar innovations used to debone red meat), chicken was still primarily served at home and in high-end restaurants. Tyson believed adamantly that chicken, which had fallen 11 percent in price from 1970 to 1978 alone, belonged on the menu of fast-food restaurants.

Tyson knew McDonald's had the best distribution system of any fast-food franchise in the United States, so he began relentlessly cold-calling its headquarters. He managed to set meetings with buyers and executives, insisting chicken was the future. He argued that chicken could be sold at half the price of the hamburger and was just as versatile—capable of being served in sandwiches or as a boneless patty. Most importantly, he made it clear only Tyson Foods could deliver the millions of pounds of meat required for a national food chain to distribute its products. But McDonald's wasn't interested—that is, until it unveiled the Chicken McNugget.

Donald Tyson's moment had finally arrived. Under his leadership, the company repurposed a plant in Nashville, Arkansas, to make only Chicken McNuggets. Tyson created an entirely new breed of chicken, which they called "Mr. McDonald," one specifically designed to improve the nugget-making process. By 1983—just three years after it had been test-marketed—Tyson was supplying the new chicken product to McDonald's stores around the country. Within three years, Chicken McNuggets accounted for 7.5 percent of McDonald's domestic sales. As Donald Tyson had envisioned, chicken products gradually made their way to the menus of other fast-food restaurant brands across the country. By 2010, chicken overtook beef and pork as the most consumed meat in the United States. From an average of thirty-four pounds of chicken consumed annually per person in 1982 (before Chicken McNuggets hit the wider market), today people consume nearly twice as much. Though fast-food restaurants are still famous for their burgers, chicken is firmly established as one of the most marketable meats in the industry.

McDonald's started with two brothers trying to get hamburgers from grill to customer quicker and cheaper, but today it operates over thirty-five thousand restaurants worldwide and opens approximately two thousand new ones each year. McDonald's feeds sixty-eight million people per day—nearly 1 percent of the world's population. To keep it that way, McDonald's spends about a billion dollars on advertising each year, much of it directed at children—a 2013 study found that elementary school children saw an average of 254 McDonald's ads per year. No wonder it is the nation's largest purchaser of beef and pork, and the second largest purchaser of chicken, after KFC.

The obvious truth is, McDonald's is only one of many fast-food restaurants delighting the millions of convenience-craving people who buy items from their "meat-sweet" menus. A quarter of Americans today eat at least one fast-food meal every single day. From several thousand in 1950 to seventy thousand in 1970, there are now over two hundred thousand fast-food restaurants located across the United States; globally, there are nearly one million. Each major chain—and there are hundreds of them—has its own unique origin story with its own colorful cast of entrepreneurial characters, but all give insight into how twentieth-century innovations in convenience gradually made their way into nearly every aspect of the fast-food empire.

II

WHY WE'RE STILL HOOKED ON MEAT TODAY

7

Death by Bliss

I hate eating vegetables. The only vegetables I eat are lettuce on a burger. —*Chance the Rapper*

"Have you ever eaten an avocado?" I asked my mom.

She replied "no" in a tone suggesting she was mildly embarrassed. My dad admitted the same, adding, "I think I've heard of them before."

At first, I was dumbfounded. "How in the world is that possible?" I replied after dramatically smacking my hand against my forehead. But, after a moment of reflection, I realized it made some sense. My parents were among the nine in ten Americans who don't eat the recommended amount of fruits and vegetables. Avocadoes might show up at Mexican restaurants on Staten Island, but my parents only ate at American, Italian, and Chinese ones. Believe it or not, I had never even tried Indian or Thai until my early twenties, when my wife—who was born in a chic part of Los Angeles—exposed me to a whole new culinary world in Manhattan. Given they had never had a meal with avocado in it before, how would my parents react to eating one?

I was about to find out. On this particular visit to my parents' house in Staten Island, I had brought guacamole with me. After explaining what ingredients are typically found in guacamole—avocado, onion, tomato, lime juice, and cilantro—I rose from the table, grabbed a small container of it from the refrigerator, and placed it in the middle of the table. I added some chips to a bowl and placed it next to the guacamole. My mom looked skeptically at the green concoction and mumbled something about not being able to eat it because she had just had work done on her teeth. My dad joked that if his food wasn't "brown like meat," he wouldn't touch it. I took a finger, skimmed off a morsel of guacamole, and popped it in my mouth. My mom, seeing that I didn't keel over, hesitantly did the same. Her reaction was the exact opposite of mine. Immediately her nose wrinkled, her eyebrows lowered, her eyes narrowed, and her tongue protruded. Finally, she managed to utter, "*Ughhh.*" Dad's reaction was less intense, but similar. "I don't like it," he said flatly. "I like the chips though."

I was disappointed. In retrospect I would have been happier if my parents had swung wildly from positive and negative reactions, like the woman who drank kombucha on TikTok, but no—their verdict was simply "ew." I was so struck by the fact that different people could have such dramatically different reactions to the taste of the same food. Taste clearly lies in the beholder of the tongue that experiences it. As Gandhi wrote in his autobiography, "The real seat of taste was not the tongue but the mind."[1] But what factors determine whether food will delight or disappoint? Might they help explain why most people are slow to reach for fruits and veggies but gleefully pile on the meat?

Salt, sugar, and fat. Few other words have occupied the minds of public health experts for the past half-century with such potency. As fast-food restaurants came into prominence and the processed food industry blossomed, there was an explosion of foods that were squeezed into them, layered on top of them, or both. At various times, salt, sugar, and fat have each been demonized as a modern plague, serving as a transient pat answer to the question: What is killing us? Though we may never know just how unhealthy they are, nearly every nutritionist agrees we are eating more than we should of all three.

The average American consumes nearly twenty teaspoons (or eighty grams) of sugar every day. The 2015–2020 Dietary Guidelines for Americans recommend limiting added sugars to no more than 10 percent of total calories. The World Health Organization actually recommends half of that. Among American adolescents, 88 percent consume more sugar than is recommended. As for salt, the average American consumes about thirty-four hundred milligrams of sodium per day, while the Guidelines recommend daily intake be no more than twenty-three hundred milligrams. Finally, the American Heart Association suggests not consuming more than 5 or 6 percent of calories from saturated fat. That's about thirteen grams of saturated fat per day on a two-thousand-calorie diet. The average American eats double the recommended amount.

The simple reason we overconsume salt, sugar, and fat is because they delight our taste buds. This is because they are essential for survival, and we evolved to seek out foods containing them. Conversely, we tend to avoid very bitter and sour foods as a defense mechanism to detect potentially harmful toxins in plants and under-ripe fruit, rotten meat, and other spoiled foods, respectively. Without supermarkets or safari buffets, our ancestors had to work hard to extract salt from the earth (it was easy while living in the ocean but difficult on the hot and dry land) and find elusive sugar-rich fruits and fatty game. The problem today, of course, is most people have unlimited access to them in increasingly processed forms.

Let's take a moment to understand how taste works: You place a piece of food (a chicken nugget, say) into your mouth. Your teeth crush the large piece into smaller ones. Saliva in your mouth simultaneously moistens the food, and the enzymes in the saliva break down some macronutrients—carbohydrates into simple sugars such as maltose and dextrin, for example—and release molecules of substances that trigger taste. These molecules collide with and bind to specific receptor cells on the taste buds on your tongue (as well as on the roof of your mouth and the back of your throat). About fifty to one hundred receptor cells, plus basal and supporting cells, make up one taste bud. The average person has between two to four thousand taste buds, and they're replaced every two weeks or so. Taste buds themselves are contained in goblet-shaped fleshy projections—the small bumps dotting your tongue. They are programmed by a few dozen genes, and each is capable of teasing out critical information from a biochemical bonanza. To do so, an electrochemical chain reaction begins in which your cranial nerves carry the message all the way to your brain. The brain interprets the message as one of five basic tastes—sweet, salty, bitter, sour, and umami (savoriness). While these are the classically depicted basic tastes, a body of research suggests fat belongs on the list. When it comes to fat, scientists know fatty acids—the building blocks of foods like oil, butter, cheese, and meat—are the stimuli. And scientists also know we have taste receptors for these fatty acids in our mouths. But researchers have yet to identify exactly how the receptors on our tongues signal the presence of fat to our brains, and that gap in knowledge disqualifies fat from joining the ranks of the other five for now. The bottom line, however, is that salt, sugar, and fat make our brains very happy.

To say salt, sugar, and fat fill us with joy is an understatement. Research suggests they are downright addictive. I don't mean "addictive" like popping bubble wrap or Netflix bingeing; I mean they are *biologically* addictive—to the point that some people argue they should be regulated like alcohol and tobacco. How exactly are they addictive?

Like with many pleasurable activities—think sex or taking certain drugs—eating can also trigger the release of the "feel-good" neurotransmitter dopamine, increasing the chance the activity becomes habitual and therefore hard to give up. Rats given access to high-fat foods show some of the same characteristics as animals hooked on cocaine and heroin; they even find it hard to quit when given electric shocks. A separate study found these addictive drugs activate the same brain connections and nerve cells associated with salt cravings. This study, also in rats, showed that blocking pathways in the brain that relate to addiction could significantly diminish salty desires. Sugar, meanwhile, has been shown to trigger regions of the brain associated with reward and craving systems crucially involved in substance abuse and dependence. When you mix salt, sugar, and fat together, the addictive effects are compounded.

The result, as David A. Kessler discusses in *The End of Overeating: Taking Control of the Insatiable American Appetite*, isn't just that people preferentially choose these foods, but that people simply eat more of them. While we have larger portion sizes and more fast-food restaurants, increased availability doesn't fully explain what's been driving us to overeat. Kessler lays the blame on foods high in salt, sugar, and fat, for their "palatability" and "reinforcement." Palatability means they engage the full range of our senses, and reinforcement means they stimulate the appetite and activate the brain's reward centers and keep us coming back for more. "And it's that stimulation, or the anticipation of that stimulation, rather than genuine hunger that makes us put food into our mouths long after our caloric needs are satisfied,"[2] Kessler explains. Big Food has spent almost a century intentionally distorting the American diet with exactly that in mind. As Michael Moss explains in his bestselling book *Salt Sugar Fat*, food manufacturers specifically formulate their foods to get us to eat more and more, to the point of addiction (the topic of his latest book, *Hooked*).

Consider the case of Howard Moskowitz, a psychologist hired by the US Army to find a way to get active-duty soldiers to eat more rations. The problem was soldiers found rations boring—optimized not so much for taste but for a long shelf life even at scorching temperatures—and would discard them, not getting enough of the calories they needed. In conversations with soldiers, Moskowitz learned they "liked flavorful foods like turkey tetrazzini, but only at first; they quickly grew tired of them. On the other hand, mundane foods like white bread would never get them too excited, but they could eat lots and lots of it without feeling they'd had enough."[3] The challenge for Moskowitz was to develop food that was alluring enough to encourage consumption but not so alluring that it quickly triggered satiation. Moskowitz set his sights on salt and sugar, both of which increased the taste of food up until a certain point before quickly falling off.

Moskowitz wasn't the first scientist to notice this "inverted U" phenomenon, but he did have the vision to recognize and capitalize on its financial potential. The sweet spot of that inverted U was at the very top of it—what researchers refer to as the bliss point, or the amount of a high-reward ingredient like salt, sugar, or fat creating maximum tastiness. By the 1980s Moskowitz was a bigwig of taste optimization. He later helped General Foods develop successful formulas for cereal, desserts, and coffee; turned Vlasic's flopping pickles into winners; and beat the competition with Campbell's line of Prego spaghetti sauces. In addition to persuading these companies to view American consumers not as monolithic, but as distinct sub-groups—lovers of traditional, more chunky, and garlicy, for example, as it pertains to tomato sauces—with different bliss points.

The prevalence of salty, sweet, and fatty foods is a major reason why people turn their nose up at fruits and vegetables and embrace processed foods, including meaty ones like bologna, bacon, ground beef, sausages, and chicken nuggets. But they don't fully explain our love affair with meat. For the final piece of the puzzle, we need to consider the role of umami.

It all began in 1907 with a chemist at Tokyo Imperial University named Kikunae Ikeda. As someone who ate a lot of dashi, a clear stock essential to Japanese dishes like miso soups and noodle bowls, Ikeda wondered what gave it—and foods like meat, cheese, tomatoes, and mushrooms—such a unique taste. Over the course of a year, he took broth distilled from dried kombu and extracted a substance called monosodium glutamate (MSG), an edible form of glutamic acid. Ikeda was convinced monosodium glutamate (and glutamate more broadly) was responsible for the taste. He tentatively named the taste "umami," borrowed from an apt combination of the Japanese words *umai* (delicious or pleasant) and *mi* (nature or taste). Ikeda never meant for "umami" to be the permanent name for his discovery, but instead suggested it as a temporary name until something better could be found. Despite his humble intentions, the naming stuck.

It took Western scientists nearly a century to acknowledge umami as a basic taste (perhaps the same vindication eventually awaits fat) since most believed it was merely a combination of the other four. It wasn't until 2000 that researchers isolated umami receptors that detect glutamate on the human tongue, granting the historically slighted taste the status Ikeda knew it deserved all along. Scientists now know it's when present with other tastes that umami's savoriness really comes alive. In nature, three umami compounds occur: glutamate (glutamic acid to which a mineral ion is attached), disodium inosinate (IMP), and disodium guanylate (GMP). Many umami-rich foods like bacon, anchovies, and cheese are also salty, and that's not a coincidence. On its own, glutamate triggers a subtle taste. But when the sodium in compounds such as MSG, IMP, and GMP interact with it, the umami taste is amplified. Umami substances have unique tastes on their own, but show synergism; when two umami substances are mixed, the mixture effect is stronger than the sum of the effects given by the two distinct substances. That's one big reason a cheeseburger with mushrooms tastes so delicious; it contains glutamate, GMP, and IMP—what some chefs refer to as "umami bombs."

The year after Ikeda began to extract MSG, he applied for and received a patent for "a manufacturing method for seasoning with glutamic acid as the key component." Ikeda contracted Saburosuke Suzuki II, then head of Suzuki Pharmaceutical Company, to produce and market the seasoning. Suzuki picked Ajinomoto for the brand name, which means "quintessence of flavor" in Japanese, and his production and distribution system for it evolved into the Ajinomoto Company. Japanese consumers could purchase it by 1909, and in the following decades, Chinese and American consumers could as well. By the mid-twentieth century, many other companies were deploying a variety of umami promoters in their products, including hydrolyzed protein, autolyzed yeast, and torula yeast, along with IMP and GMP. The United States produced about fifty-eight million pounds of MSG in 1969, much of which was added to premade products such as TV dinners and canned soup, rather than sold directly to consumers as a stand-alone condiment. Today, Americans consume nearly three times as much as they did in 1969 (about 500 milligrams per person per day) and it is commonly used by manufacturers and fast-food restaurants like KFC and Chick-fil-A.

Why does umami taste so delicious? Just as sweet foods signal the presence of energy-packed sugar and salty foods indicate the presence of minerals, research suggests umami-rich foods alert us to the presence of protein, a macronutrient essential to survival. Whatever the evolutionary context for the origins of our ability to experience umami, we now know from neuroimaging studies that glutamate stimulates reward circuits of the brain similar to other taste-inducing substances. We are even exposed to glutamate in utero, as it can be found in the amniotic fluid. Once we're born, we take in glutamate at high levels during breastfeeding; human breast milk has ten times more than cow's milk has, in fact.

But taste is only one relatively minor component of the sensory eating experience. According to Barb Stuckey in *Taste What You're Missing: The Passionate Eater's Guide to Why Good Food Tastes Good*, "Some [experts] say that only about 5 percent of what we experience when eating is input from our sense of taste. They think that the remaining sensory input—the vast majority—is aroma, which we detect with our nose."[4] To experience what they mean, close your eyes and plug your nose while you sprinkle salt on your tongue. You'll taste it immediately. But salt doesn't have a flavor. Without your nose, you'll be able to taste, say, bacon, as salty, but you won't recognize it without its aroma. In addition to dosing your taste buds, chewing food also releases aromatic gases that float up into the nasal cavity, which connects with the back of your throat and is directly above the roof of your mouth. On the roof of the nasal cavity is the olfactory epithelium, which contains specialized receptors that respond to certain odor molecules. When the smell receptors are stimulated, a signal travels along the olfactory nerve to the olfactory bulb, situated underneath the front of your brain just above the nasal cavity. Signals are then sent from the olfactory bulb to other parts of the brain to be interpreted as a smell you may recognize, like beef jerky or tuna tartar. There are hundreds of olfactory receptors, each capable of sorting out subtle and complex smells. Together, taste and smell primarily form the unique multisensorial experience known as flavor.

There are two classes of flavor—natural and artificial—and both are exploited by companies to make their products as addicting as possible. "Natural flavor" indicates the molecules in the flavoring came from ingredients found in plants or animals, though they don't necessarily come from the same food as the target flavor; castoreum, for example, is completely natural because it's secreted by glands located in the butt of a beaver, and compounds derived from it can be used as a natural vanilla or raspberry-like flavoring. Conversely, "artificial flavor" means the molecules were not extracted from nature but rather created in a laboratory. For example, many fruits—for example, apples, oranges, strawberries, and mangoes—get their flavor and aroma from specific chemical compounds called esters (which, by the way, are also used in some perfumes and cosmetics). By synthesizing these chemicals using various organic compounds, such as carboxylic acids and alcohols, flavor scientists can create artificial flavors that are identical to the real thing. In other words, the distinction in flavorings—natural versus artificial—comes from the source of these chemicals, not the structural composition.

As Mark Schatzker explains in his book *The Dorito Effect,* flavor technology got very powerful in the 1960s: "One after another, humans have captured the chemicals that characterize foods like apples, cherries, carrots, and beef and moved their production from plants and animals to factories. In 1965, there were less than 700 of these chemicals. Today, there are more than 2,220."[5] Schatzker further explains that flavor scientists would have been unable to create nearly all of them without a single invention: gas chromatography. Though precursors of it were developed in the mid-nineteenth century, modern gas chromatography allowed chemists to identify and separate the individual compounds (based on characteristics like their unique boiling points and polarity) that gave a substance its unique flavor. The first commercial gas chromatography unit became available in 1955, and with it, scientists became food artists, shaping and crafting various molecules to achieve a desired flavor.

Journalist Melanie Warner paints a fascinating visual of this process in her book *Pandora's Lunchbox:* "To do this, they trek to farms when a crop is at the peak of ripeness, taking portable gas chromatography devices with them. They drape the strawberry or tomato or pepper plants with plastic bags or glass jars, corralling the aroma gasses in an attempt to make an imprint. . . . Back in the lab, you work on mimicking what the machinery has identified."[6] For example, delta-dodecalactone was found in butter and added to margarine, while cis-3-Hexen-1-ol, also known as leaf alcohol because of its intense odor reminiscent of freshly cut green grass and leaves, was found in plants and added to all types of fruits. The list goes on and on. The motivation is simple: The flavoring industry makes billions annually by closing deals for its flavorings with food companies (many of which also rely on taste enhancers like sugar and glutamate). As explained by a senior flavorist at Givaudan, the biggest "flavor house" in the world, because so many products are heavily processed, they lose flavor and so food scientists must "add back to that and make that taste like Mother Nature intended."[7]

Meat isn't immune to the wizardry of flavor enhancement. In 1989, for example, Ajinomoto Company (the MSG purveyor) found another magical ingredient. Scientists at the company discovered when water extract of garlic was added to Chinese soup and curry soup, it both heightened the meaty flavor and increased "continuity, mouthfulness, and thickness"[8] despite not having any taste of its own. To find the key compounds which gave rise to the effect, the water extract was chromatographed, and a protein called glutathione was identified. Ajinomoto Company referred to this effect, in which compounds in food that don't have their own flavor enhance the flavors with which they're combined, as "kokumi," sometimes translated to English as "heartiness" or "mouthfulness." Since then, food scientists have introduced into meat other kokumi-eliciting compounds including calcium, protamine, yeast extract, and L-histidine. According to Schatzker, "If you have eaten chicken wings in a restaurant lately . . . there is a good chance a billion-dollar multinational you've never heard of stimulated your kokumi receptor."[9] Take a peek at the ingredients used to make a McDonald's chicken nugget in the United States: white boneless chicken, water, vegetable oil (canola oil, corn oil, soybean oil, hydrogenated soybean oil), enriched flour (bleached wheat flour, niacin, reduced iron, thiamine mononitrate, riboflavin, folic acid), bleached wheat flour, yellow corn flour, vegetable starch (modified corn, wheat, rice, pea, corn), salt, leavening (baking soda, sodium aluminum phosphate, sodium acid pyrophosphate, calcium lactate, monocalcium phosphate), spices, yeast extract, lemon juice solids, dextrose, natural flavors. Though, according to McDonald's, it is "made with 100 percent white meat chicken and no artificial colors, flavors or preservatives,"[10] a chicken nugget is a lot more than chicken.

Here's the rub: As food anthropologist Bee Wilson explained to me, the more we eat foods that are rich in taste and flavor enhancers, the more we crave them: "Lots of us don't think that we do learn to eat. We think of eating as some kind of innate thing that we are born knowing how to do like breathing. And that's not the case. It turns out that actually eating is a series of really complicated skills that we learn from babyhood onwards. But the main way we learn how to eat is through exposure. Scientists have known for a very long time that most of what we think of as our likes and dislikes are a function of the particular flavors we've been exposed to."[11]

Social psychologists refer to this phenomenon as the "mere-exposure effect" (also known as the familiarity principle), a term coined by social psychologist Robert Zajonc in 1968 to describe people's tendency to develop a preference for things they are familiar with. Acquired tastes for bitter foods like coffee, wine, and beer intuitively suggest that some tastes can change with repeat exposure, but research supports it too. In a seminal study, Zajonc showed subjects a series of random shapes in quick succession. The shapes flashed so rapidly it was impossible to discern that some were actually repeated. Nevertheless, when subjects were later asked which shapes they found most pleasing, they consistently chose the ones to which they had been exposed the most often despite not being consciously aware of them.

Studies have continued to look into the effects of mere-exposure on edible substances. For the most part, it turns out we dislike the flavor of what we don't know and grow to like the flavor of what we do know. In a 2009 study, children aged between twenty-one and twenty-four months received daily exposure to one of two books. The books featured two foods that were familiar to them, and two foods that were unfamiliar to them. After two weeks of reading their assigned books, the children participated in a taste test. Each child had been exposed via their books to two of the four foods they were served. Sure enough, they expressed a willingness to taste significantly more of the foods that were familiar to them. Studies have since continued to confirm the link between familiarity and taste preferences.

In contrast to exposing children to fruits and vegetables, as Bee Wilson explains in her book *First Bite: How We Learn to Eat*, "It is clear that large numbers of adults as well as children have now become habituated to eating a version of 'kid food' over a whole lifetime: sweet, salty, undemanding to chew and swallow, and heavily processed. The kind of menus you typically see at casual chain restaurants suggests that when adults go out to eat, they want childish comfort: sweet-salty ribs, breadcrumbed chicken, cheesy pasta."[12]

In other words, foods more commonly associated with the kid's menu are now giving adults increasingly immature palettes. This might explain why my parents hadn't eaten an avocado until their sixties, and why they didn't like the taste. By barraging their taste buds with products concentrated in taste-triggers and flavorings formulated for maximum "bliss," foods like guacamole taste bland or overwhelming by comparison.

But scientists aren't only juicing meat with taste-enhancers and flavor concoctions; countless processed plant-based foods have them too. This begs the question: What is so special about the flavor of meat that keeps us hooked?

Almost all roads to this culinary question first lead to a French physician and chemist named Louis Camille Maillard. Maillard rose to prominence thanks to his work not on meat but on the metabolism of urea and kidney disorders. He initially discovered that heating sugar and amino acids together in a test tube turned the mixture brown. Not until World War II did anyone notice a connection between the Maillard reaction and flavor, when soldiers were complaining about brown food that didn't taste very good. Scientists eventually determined the unappetizing tastes and color changes were due to the Maillard reaction. At first, they went to work trying to prevent it, but they soon saw its potential as a flavor *enhancer*.

Though the underlying chemistry involved in rearranging amino acids and simple sugars is, as you might imagine, quite complex, the result is simple and creates a unique "meaty" flavor we find delicious. The flavor is born from over a thousand new molecules generated during the Maillard reaction, which usually occurs above 285°F; without it, meat will have less flavor. That's one of the main reasons why we—and many other animals like mice and chimps, as demonstrated in various experiments—will consistently choose cooked over raw meat when given the option. While it's true the Maillard reaction plays a role in enhancing the overall flavor of other foods, such as popcorn, chocolate, and beer, it is arguably most potent in meat. The Maillard reaction keeps us hooked on tender steaks, juicy grilled hamburgers, and deep-fried chicken wings, and it's also a reason these foods can be very unhealthy, particularly in excess, potentially causing dangerous compounds to form, including probable carcinogens like acrylamide and furans.

Yet another reason we find the flavor of meat so delicious is because it contains a unique combination of naturally occurring fats. While it is true that many factors have been found to influence the aromas of cooked meat, researchers have found that among all food constituents, fats typically have the greatest influence on the production of them. This makes sense from a human evolution standpoint, as fats have more than twice the number of calories per gram compared to proteins and carbohydrates. Although the composition of these fats and subsequently the taste varies widely depending on many factors, most notably the source of the meat (cow, chicken, pig, etc.) and the diet the animal ate (grain vs. grass fed, for example), the result is the same. When you cook a piece of meat, in addition to the Maillard reaction, fats start to oxidize, releasing many delicious aromas.

In addition to being chock-full of naturally occurring glutamates, there are countless other small-part players in the overall flavor of meat. Together, they coalesce to form a sensory experience that delights our palates in ways no other food seemingly can. When loaded with salt, sugar, fat, and umami triggers, as well as various flavorings—whether extracted from nature or formulated in a lab—meat becomes even more irresistible.

But industrial animal agriculture also knows tastes can come and go. Turtles, beavers, and eels were once beloved staples of the continental diet, but now they are in the culinary dustbin of history. Could this one day happen to meat altogether? To ensure its products don't become another fleeting food fad, industrial animal agriculture's tricks have to go beyond taste manipulation.

8

Tricks of the Trade

One day in the fall of 2018, I was mindlessly scrolling through my Facebook newsfeed when I came across a peculiar post on *Quartz*. It contained an idealized image of a brawny farmer in a cowboy hat riding a horse, and an accompanying article with the headline: "There's an Answer to Cattle's Carbon Emissions—and It Isn't Less Beef." Out of skepticism and morbid curiosity, I clicked, as we often do. Here's a sample:

> Cattle production has long been a bedrock of American agriculture, and currently, there are more than 913,000 cattle ranches and farms operating around the country, comprising more than 40% of US farms.... Beef supply chains are increasingly globally interconnected, especially as the number of imports and exports rise year-over-year. The livestock industry as a whole provides the livelihoods for more than a billion people around the world, and contributes to more than 33% of a global protein consumption. However, as concern about climate change grows, interest in quantifying and mitigating emissions from global livestock production has also increased. The entire global livestock industry creates roughly 14.5% of greenhouse gas emissions, and global beef production is responsible for 6% of emissions. And while it's clear that better efficiencies are needed to alleviate the worst environmental effects, it's also important to understand

the ways the beef industry positively contributes to the environment and human welfare. The data surrounding the issue can help us fund solutions.[1]

The "data," presented in colorful, interactive charts and maps alongside idyllic images of wheat shimmering in the sun and cows happily foraging the open land led to a single conclusion: "The rest of the world hasn't kept up with the growing efficiency of American beef production." If it had, the article goes on to explain, "carbon emissions would be much lower." The authors make two recommendations: (1) support "the widespread American practice of grain-finishing beef" and (2) increase "the efficiency of agricultural production, often referred to as 'sustainable intensification.'"

I was amazed *Quartz* staked its reputation on an article with such dubious conclusions, but then I looked deeper. The article was "sponsored," meaning someone paid for it to appear under the *Quartz* banner. Sure enough, at the bottom of the page, was this disclaimer: "This article was produced on behalf of the National Cattlemen's Beef Association."

The National Cattlemen's Beef Association (NCBA) is a trade association for beef producers in the United States. It's one of many organizations, like the National Chicken Council (NCC), the National Pork Board, and the American Association of Meat Processors (AAMP), that protect the interests of the meat industry. But exactly what tactics do they—as well as food companies in the industrial animal agriculture complex—use to lure us back to the trough again and again?

You've almost certainly heard another one of industrial animal agriculture's marketing campaigns: "Beef. It's What's for Dinner." Surveys show that nearly 90 percent of American adults know it, making it one of the most successful slogans in advertising history. The Beef Industry Council unveiled the campaign in 1992, and the admittedly catchy tagline (which somehow overcomes the clunky use of two contractions in a row) was primarily blasted through television and radio advertisements. They're narrated by actor Robert Mitchum, who touts the diverse and savory beef entrees from around the country accompanied by triumphant country music. The first campaign ran for just under two years and cost $42 million. The ads made a comeback in 2017, albeit in a new form to keep up with the changing times. In between, it was replaced by a less successful but still impactful campaign and slogan: "Beef. It's what you want," and

"Discover the Power of Protein in the Land of Lean Beef." As Alexandra Bruell writes in the *Wall Street Journal*, "The social-media campaign from the National Cattlemen's Beef Association combines nostalgic elements, such as the tagline and narration that alludes to the 'Old MacDonald Had a Farm' nursery rhyme, with a more modern story line about how beef farmers and ranchers are using technology."[2] Bruell notes the Association's potential marketing budget for the resurrected campaign including research, promotions, and advertising is about $30 million. So where did that money come from, exactly?

The programs responsible for creating this award-winning advertising campaign—along with many campaigns of similar ilk, like "Pork. The Other White Meat," and "American Lamb from American Land"—are known as "checkoffs." You may also be familiar with "The Incredible, Edible Egg," "Ahh, the Power of Cheese," "Got Milk?," the "Milk Mustache" campaign and "Milk: It Does a Body Good"—all funded by checkoff programs. You can think of checkoffs as like a mandatory tax on products sold. Unlike an advertisement for a specific company or brand (say, Tyson Foods), the messaging is generic, a result of a cooperative effort of a large group of producers to promote products with similar characteristics—in this case, meat, dairy, and eggs. Farmers are required to pay a certain percentage of their harvest into a national fund. The current beef checkoff, for instance, requires ranchers to pay $1 per head of cattle into the fund; the current pork checkoff requires producers to pay $0.40 per $100 of value; and the current lamb checkoff requires producers and feeders to pay an assessment of $.007 per pound. A board comprised of industry stakeholders, appointed by the secretary of agriculture and supervised by the Agricultural Marketing Service, the body of the USDA that oversees the twenty-two checkoff programs, then allocates a portion of the proceeds among state and regional industry organizations and decides how to spend the funds on research, education, and marketing.

How effective are checkoff programs? Though estimates vary widely, one of the most comprehensive studies ever rendered on checkoffs concluded each dollar invested in the beef checkoff program between 2006 and 2013 returned about $11.20 to the beef industry; the report noted in its absence, total domestic and foreign beef demand would have been 11.3 and 6.4 percent less, respectively, than it was with the checkoff programs in place. The National Pork Board notes an even higher return, citing on its website that in "the 2017 Return on Investment analysis of the Pork Checkoff, $25 of producer value was returned for every $1 invested."[3]

This may seem like an outrageously good investment, but many independent animal farmers point out the gains are not shared equally. "The checkoff doesn't help me," said Rodney Skalbeck, a pig farmer in Minnesota, in *In Motion* magazine. "It goes to help sell pork, but packers and retailers just keep that money, and the farmer gets less and less. It's clear that the pork checkoff is a failure."[4] In an editorial for the *Des Moines Register*, farmer Chris Petersen writes, "We have lost 94 percent of traditional hog farmers in this country. These farmers have been squeezed out by industrial agriculture and by policy—and checkoff programs—that favor factory farms over our nation's independent food producers."[5] Just four companies produce about 85 percent of US beef. A campaign promoting beef does nothing to encourage consumers to avoid beef from these multinational corporations that rely on factory farms in favor of beef from independent farmers raising grass-fed cattle,* which is likely to be marginally costlier. This is what prompted a coalition of independent hog farmers named the Campaign for Family Farms to try to abolish the pork checkoff program. They put forth a referendum in 2000 in which over thirty thousand hog farmers voted to end the mandatory pork checkoff program by 53 percent to 47 percent. But the then new secretary of agriculture, Ann M. Veneman, overturned the results on the grounds that the petitioning had been procedurally flawed. The *New York Times* editorial board published a scathing column in support of halting the checkoff programs on the grounds that it "no longer makes sense for reasons that have to do with the seismic shift in the hog industry. In 1985, when the Pork Act was passed, some fifty-two million pigs were raised on more than 388,000 farms in America. Last year [in 2001], nearly sixty million pigs were raised on fewer than 82,000 farms. These days the industry is dominated by factory farms, and the family farmer who raises hogs is an exception."[6]

---

* Grass-fed is actually a misnomer. All cows raised for food—even factory-farmed ones—spend a little over a year on the pasture. The final six to eight months for conventionally raised cattle are spent on a feedlot, where they are finished on a grain-heavy diet to promote maximum weight gain. "Grass-finished" cattle remain on the pasture during the finishing process.

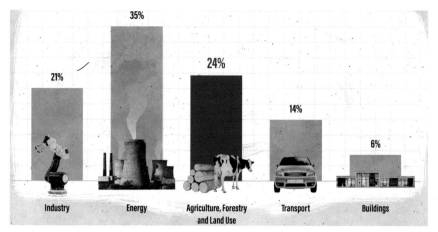

Agriculture, especially animal agriculture, is a major contributor to greenhouse gas emissions.

Helen Atthowe of Woodleaf Farm in Eastern Oregon is one of a handful of farmers in the nation who embrace veganic, regenerative practices.

Cultured meat cells grow in a specially formulated media— typically a mix of nutrients, salts, pH buffers, and growth factors that allow the cells to proliferate.

It turns out that it's hard to change people's attitudes and behaviors. I learned this the hard way as a teenager when I tried convincing my parents to recycle by making them a special pail. My attempts to get them to decrease their meat consumption have gone about as well.

Believe it or not, this was the first time my parents tasted guacamole. It went about as well as I expected.

Will Harris gives me a tour of his farm, White Oak Pastures, a sixth-generation, 152-year-old family farm in Bluffton, Georgia.

Tobey, my dog, is part of a long history of domesticated animals, arguably no different than cows, pigs, chickens, and sheep.

At the "pig vigil," I caught a glimpse of two pigs on their way to slaughter.

A scientist at Beyond Meat shows me how they isolate and study flavors from animal-based meat in an effort to replicate them using plant-based ingredients.

At long last I taste Eat Just's cell-cultured chicken nugget.

Pigs on Will Harris's farm have won the jackpot— they are allowed to roam free and express their instinctive behavior. Pigs raised on industrial farms aren't nearly so lucky.

Parker Lee, lead scientist of Beyond Meat's analytical lab, puts my nose to the test, asking me to identify some of the signature aroma molecules of meat.

If taste is king, texture is queen. Beyond Meat and other plant-based meat companies use specialized machinery, like this "e-mouth," to achieve a mouthfeel that matches its animal-based counterparts.

Carl's Jr. is one of many fast-food restaurants offering plant-based burgers, capitalizing on the three primary motivators in decision-making around food: price, taste, and convenience.

Staten Island isn't exactly a plant-based food Mecca, but its many parks and wildlife turned me into a card-carrying environmentalist.

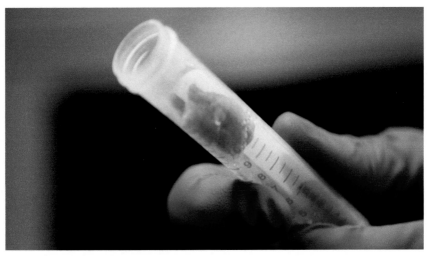

Cell-cultured meat is prototyped in a lab, but it's grown at scale in a facility that looks more like a beer brewery.

Many plant-based meat companies use ingredients from nature to create a blood-like color. Beyond Meat, for example, uses beets.

Plant-based burgers from the alternative protein company Beyond Meat require 93 percent less land and 99 percent less water compared to animal-based burgers.

The voters in the majority were not happy, to put it mildly: "Secretary Veneman has turned her back on us," said Mark McDowell, a pig farmer in Iowa. "Thousands of pork producers cast their ballots knowing that the program would end within 30 days of the vote announcement, if a majority voted to terminate, and that is exactly what happened. Veneman's inaction means we have to continue paying $1 million a week. We do not support the pork checkoff tax. Veneman needs to terminate the program now." Ultimately, a settlement agreement allowed the mandatory program to carry on. And it has ever since.

While it's true there are checkoffs for all sorts of commodities, from cotton to watermelons, the animal agriculture industry as a whole benefits disproportionally. Of the $850 million generated from checkoff programs annually, nearly half goes to the National Dairy Promotion & Research Board, and approximately $100 million goes to lamb, beef, and pork checkoffs. The few plant foods, meanwhile, represented by checkoffs include almonds, mushrooms, mangoes, sorghum, watermelons, peanuts, raspberries, popcorn, blueberries, potatoes, and soybeans (which is part of industrial animal agriculture, as over 70 percent of the soybeans grown in the United States are used for animal feed). Everything else, from chickpeas to tomatoes to broccoli, must be supported by private sector funds, not government programs. With few exceptions (the Hass Avocado Board has a budget of over $50 million, for example), each has such a small budget that it can't pay for the kind of marketing that can boost its consumption. The National Mango Board, the Mushroom Council, Highbush Blueberry Council, and the National Watermelon Promotion Board, for example, collectively have a budget about a quarter of the size of those that represent industrial animal agriculture. So why doesn't the national fruit and vegetable industry simply unite by forming its own checkoff? Well, one that would raise approximately $30 million through a 0.046 percent assessment from first handlers and importers was actually proposed in 2009, but the farmers ultimately rejected it in part because they worried the sum would be a drop in the bucket when compared to the meat, egg, and dairy boards' $500 million-plus.

Checkoff funds are meant to be politically neutral, existing only for research, promotion, and education purposes. But without adequate oversight, the lines are sometimes blurred, and they may be used questionably to influence government policy. For example, in August 2021, the Beef Checkoff ran a full-page ad in the *Wall Street Journal* claiming that "if all U.S. livestock were eliminated and every American followed a vegan diet, greenhouse gas emissions would only be reduced by 2 percent, or 0.36 percent globally."[7] The statistics were misleading on the verge of preposterous, but the intent was clear: Cast doubt on well-established climate science to keep public opinion on the industry's side, dampening enthusiasm for any future government action.

Next, consider a lawsuit filed in 2012 by the Humane Society of the United States, Iowa Citizens for Community Improvement, and independent pig farmer Harvey Dillenburg against the USDA and the then agriculture secretary Tom Vilsack (who is once again serving this role in the Biden administration). It argued the National Pork Producers Council (NPPC), which ran the marketing slogan "Pork: The Other White Meat" from 1986 to 2001, has been financing its lobbying efforts in part through the sale of associated trademarks to the National Pork Board. Furthermore, it alleged the National Pork Board agreed to pay and the USDA approved a wildly inflated price of $35 million plus interest over twenty years—more than double what Steve Meyer, an agricultural economist and consultant to the National Pork Board, valued the slogan to be worth. (In fact, the slogan was retired in 2011 and replaced with "Pork: Be Inspired.") In February 2018, a federal district court in Washington, D.C., issued a ruling that while any previous payments are moot, the National Pork Board must cease further payments to the NPPC, stating "the trademarks that include The Other White Meat slogan have been declared to be obsolete, and have been retired from active use. So their value is minimal, or at best, undetermined." However, in August 2019, the ruling was overturned and the case dismissed, citing that there wasn't "any indication that the 'price for pork' was 'affected' by the alleged misuse of checkoff funds." Nonetheless, the cozy relationship between the entities that administer the checkoff funds and those that engage in lobbying is self-evident.

Ultimately, the budgets of checkoff programs are miniscule in comparison to industrial animal agriculture's overall sales and market worth. According to the North American Meat Institute, a nonprofit industry trade association, the red meat and poultry industries generate $1.02 trillion annually to the US economy, or roughly 6 percent of our entire GDP.

(By comparison, a recent report estimated the entire US fruit and vegetable market was valued at $104.7 billion.) And according to a new report by Grand View Research, Inc., the global meat market (including seafood) is expected to reach $7.3 trillion by 2025. As food journalist Deena Shanker noted in *Quartz*, "That size translates into political influence."[8] According to the Center for Responsive Politics, in 2016, the red meat and poultry industry spent approximately $11 million in contributions to political campaigns, and nearly $8 million directly on lobbying the federal government.

To see how these dollars translate to influence, consider Iowa's hog industry. Pigs in the Hawkeye State outnumber humans seven-to-one while producing the same amount of waste as 84 million people—more than California, Texas, and Illinois combined. Iowa is by far the nation's largest pork producer, raising roughly a third of the nation's hogs, and the state's largest operation is Iowa Select Farms. The farm and its owner, Jeff Hansen, have donated generously to local politicians, including $300,000 to the campaign of Iowa Governor Kim Reynolds. When the COVID-19 pandemic swept through the Midwest, Governor Reynolds, according to a sweeping *Vox* exposé, "fought to keep plants open, prioritizing farmers like Hansen, who would lose millions as barns became overloaded with market-ready animals."[9] When Iowa Select's Headquarters experienced a COVID-19 outbreak scare in July 2020, the governor immediately sent a rapid-response team to test the office's mere thirty-two employees.

According to Steve Johnson, one of the coproducers of *Modern Meat*, an in-depth documentary aired in 2002 by *Frontline*, "Instead of spreading lots of money around to many different lawmakers in an attempt to gain access and influence—the traditional method used by many large corporations—the meat industry targets their approach to a small number of key lawmakers and regulators that have a direct impact on their business interests."[10] Johnson brings up a well-cited example to demonstrate how industrial animal agriculture influences those on Capitol Hill, one involving a series of events following the 1993 Jack in the Box E. coli outbreak.

During this incident, 732 people (mostly children) were infected with E. coli originating from contaminated beef patties purchased from Jack in the Box restaurants in the western half of the United States. When the USDA proposed implementing new food safety regulations in response to the outbreak, "The [industrial] meat industry attempted to delay the implementation of the new regulations by convincing a member of the key appropriations committee to introduce an amendment to stop the rulemaking process," writes Johnson. The chosen member this time was Rep. James Walsh, an Upstate New York Republican who since 1988 has received $66,000 in campaign contributions from the meat and food industry. Walsh slipped a one-paragraph rider into a subcommittee report that was intended to halt the new food safety program for nine months, under the guise that more time was needed to study and comment on it.

At the time, inspectors in the USDA's Food Safety and Inspection Service (FSIS)—the entity tasked with the responsibility of inspecting meat and poultry—relied on the "poke and sniff" method, which is exactly what it sounds like: they would literally touch, smell, and prod the meat to distinguish contaminated meat from clean cuts, a near impossible task given that many pathogens and microbes are undetectable by scent and the naked eye. The industry's biggest objection to the proposed inspection program was that it would require more rigorous testing for exceeding limits of salmonella—a group of bacteria often found in the feces of some animals—in ground beef, which it argued was not the proper scientific measure to use. Walsh's rider, according to the *Washington Post*, was drafted by Philip Olsson, then a lobbyist for the National Meat Association. After a public outcry from consumer advocates and newspaper editorials criticizing the amendment, Walsh and the then agriculture secretary, Dan Glickman, reached a compromise: In exchange for dropping the amendment and thus allowing salmonella testing to move forward, Glickman would allow the meat industry to air its concerns at open public hearings. In 1998, the government finally unveiled a radically redesigned system of meat inspection called Pathogen Reduction/Hazard Analysis and Critical Control Point Systems (PR/HACCP). Originally developed by NASA to ensure the safety of astronauts' food in flight, the PR/HACCP program was a science-based approach to food safety that included comprehensive microbial sampling and analysis. Health advocates rejoiced, but industrial animal agriculture wasn't ready to throw in the towel.

Its opportunity for a rematch came via Supreme Beef Processors, Inc., a Texas-based meat processor. In December 1999, one of its plants failed the PR/HACCP's salmonella tests three times in eight months, and in one test 47 percent of its product was contaminated. At this time, the company was selling millions of pounds of beef to the National School Lunch Program, the federally assisted meal program that provides low-cost or free lunches to children at school. Pursuant to the PR/HACCP regulations, the FSIS notified the company that it would have to halt operations. As the *Dallas Observer* noted, this was "the first ground-beef processor in the United States to be threatened with closure by the USDA for failing to meet what was then a 1-year-old salmonella-testing standard."[11] Supreme Beef Processors, Inc. immediately filed a lawsuit against the USDA in federal district court, and the court subsequently granted the company a temporary restraining order preventing the government from removing the inspectors. In the lawsuit, Supreme Beef Processors, Inc. argued salmonella is naturally occurring and therefore is not an "adulterant" subject to regulation by the government; it also pointed out salmonella is killed when meat is cooked properly, and therefore did not pose an eminent risk. In 2000, a judge ruled that Supreme Beef should not be held responsible because the salmonella was already present in the meat when it arrived at the plant. The landmark ruling has hindered the agency's ability to regulate the safety of red meat and poultry ever since.

In addition to making financial contributions to and lobbying politicians, industrial animal agriculture also forms partnerships and alliances with nonprofits, often directly or indirectly funding their marketing activities. In 2011, for example, the Beef Board announced a new partnership with the American Heart Association (AHA), in which boneless top sirloin petite roast, top sirloin filet, and top sirloin kabob would display a heart-shaped stamp as part of the AHA's Heart-Check Food Certification Program. According to the AHA's website, "Food manufacturers participating in the program pay administrative fees to the American Heart Association to cover program operating expenses."[12] The AHA also has a "Heart-Check Recipe Certification Program [that] employs an annual per-recipe fee-based assessment to cover program operating expenses." As of April 2021 the National Cattlemen's Beef Association has twenty. From "Tangy Lime Grilled Beef Top Round Steak" to "Sweet and Sloppy Joes," AHA endorses them all. Sadly, this isn't an isolated example of industrial animal agriculture co-opting the nonprofit sector.

Consider the 2010 Healthy Hunger-Free Kids Act. This law mandates more plant products be served in schools. (A Harvard School of Public Health study found kids ate 16 percent more vegetables and 23 percent more fruit at lunch under the new federal standards.) In 2010, the School Nutrition Association (SNA), a professional group representing more than fifty-seven thousand cafeteria and school nutrition employees, was a big supporter of the Healthy Hunger-Free Kids Act. A few years later, however, the SNA, under new leadership, began fighting against the law. Advocates at the Environmental Working Group (EWG) pointed out that more than half of SNA's funding in 2012 came from food industry members, including membership fees and sponsorship opportunities. In 2013, according to SNA's website, its national annual conference was sponsored by a diverse array of industry stakeholders including Jennie-O Turkey Store, a subsidiary of the Hormel Foods Corporation, and the National Dairy Council; in 2014, Tyson Foods joined as a sponsor. Audrey Sanchez, founder and executive director of Balanced, an advocacy organization focused on improving the healthfulness of menus in schools, told me that for these and other industry members, there is a lot at stake: According to the USDA's Economic Research Service, in 2016 the National School Lunch Program operated in over one hundred thousand public and non-profit private schools and provided lunches to over thirty million children each school day at a cost of $13.6 billion.

The SNA made headlines again in early 2017 when it released recommendations for the USDA to scale back the federal nutrition standards, "calling for practical flexibility under federal nutrition standards to prepare healthy, appealing meals."[13] At the end of the year, the SNA praised the USDA when it decided to do exactly that, and again at the end of 2018 immediately before it was set to go into effect: "School nutrition professionals have made tremendous progress in improving student diets, but the pace and degree of menu changes under updated nutrition standards were more than some students would accept,"[14] SNA president Gay Anderson wrote in a release. "We appreciate Secretary Perdue for finding solutions to address the concerns of schools and students." (The American Heart Association, to its credit, put out a scathing critique of the decision.)

In addition to working with others to roll back inconvenient legislation, industrial animal agriculture has been remarkably successful at initiating and exploiting new laws that prevent and intimidate adversaries from criticizing it—something Oprah Winfrey and her guest Howard Lyman learned the hard way some twenty-five years ago. Winfrey's tango with industrial animal agriculture began in April 1996, when her talk show aired an episode on food safety. The segment included discussion of bovine spongiform encephalopathy, commonly known as mad cow disease, which had recently killed cattle in England. Oprah invited William Hueston from the USDA, Gary Weber from the National Cattlemen's Beef Association, and Lyman, a fourth-generation Montana cattle rancher turned vegetarian, onto her show to discuss controversial practices within the beef industry, including a process now banned in the United States called "rendering," which involves turning cow organs into feed for other cattle. Lyman said while there had been no documented cases of mad cow in the United States, it was only a matter of time. That prompted Winfrey to declare: "It has just stopped me cold from eating another burger! I'm stopped!"[15] The segment coincided with what Wayne Purcell, a Virginia Tech professor who at the time produced a newsletter on the livestock market, called "a significant and rather dramatic shock"[16] to cattle prices.

As Aman Batheja reported in her exhaustive exposé in the *Texas Tribune*, members of industrial animal agriculture were furious, as was one of their biggest cheerleaders, Texas agriculture commissioner Rick Perry. Within days, Perry wrote a letter to then Texas attorney general Dan Morales urging the state to take legal action against Oprah and Lyman under food libel laws, also known as food disparagement laws and informally as "veggie libel" laws (because they have often been used against advocates who are working to promote plant-based eating and expose the harmful consequences of excessive meat consumption). Though they vary from state to state, these statutes were designed to protect the agriculture industry from profit losses arising from false claims, but in practice they enable corporations to intimidate and prevent critics from publicly criticizing the safety of their products, regardless of merit. They continue to serve that purpose today for companies in thirteen US states: Alabama, Arizona, Colorado, Florida, Georgia, Idaho, Louisiana, Mississippi, North Dakota, Ohio, Oklahoma, South Dakota, and Texas.*

---

* Unfortunately for Winfrey and Lyman, the Texas version of a food libel law known as the False Disparagement of Perishable Food Products Act had just passed in 1995.

In December 1997, Winfrey and Lyman were sued by a group of Texas cattlemen for $10.3 million in federal district court on a charge of disparaging beef under the food libel law. Specifically, they were accused of creating a "lynch mob mentality" among members of the studio audience to produce a "scary" story about the safety of beef. Leading up to the trial, Winfrey moved the taping of her immensely popular program from her studio in Chicago to the Little Theater in Amarillo, Texas. The town was divided: Though many of the 170,000 residents proclaimed their love for the mega-celebrity, others recognized the industry was a massive employer in the region. Those in the latter category spearheaded the design and sale of bumper stickers with the crude slogan, "The only mad cow in Texas is Oprah," and T-shirts sporting her face with a red line across it. Winfrey was legally prohibited from publicly discussing the case. "We're down here in Amarillo—y'all know why," she joked on her show. In the end, five weeks after the trial began, the jury deliberated for six hours and then voted unanimously in Winfrey's and Lyman's favor, finding no evidence the duo had knowingly spread false information about the industry with the intent to disparage it. When she left the courthouse, Winfrey shouted, "Free speech not only lives, it rocks!" "I'm still off hamburgers,"[17] she added.

To date, no one has been found liable under any food disparagement lawsuit, but the highly public case was effective in reminding advocates that opening their mouths in criticism against industrial animal agriculture might come with a hefty price tag in legal bills. Just ask Walt Disney Co., the corporate parent of *ABC News*. In 2012, the network released an infamous exposé of "pink slime"—referred to by industrial animal agriculture as "finely textured beef"—present in ground beef. This prompted meat processor Beef Products, Inc. to file a lawsuit against Walt Disney Co., anchor Diane Sawyer, and reporter Jim Avila in 2012 for $1.9 billion, alleging the segment misled consumers into thinking it wasn't safe to eat, which resulted in sales plummeting. The food-libel law in South Dakota provides for triple damages against those found to have violated it, so *ABC News* was facing, potentially, nearly $6 billion in damages. Five years after the segment aired, Walt Disney Co. settled for more than $177 million, the most ever in a corporate legal case of its kind.

Food libel laws aren't the only ones that may silence potential critics. Ag-gag laws—a term coined by food journalist and former columnist for the *New York Times* Mark Bittman—refer to state laws forbidding the act of undercover filming or photography of activity on farms without the consent of their owner. (The industrial animal agriculture lobby calls them "farm protection laws.") Ag-gag laws emerged in the early 1990s in response to activists and journalists who engaged in "undercover investigations," James Bond–like stealth operations to expose wrong-doing. Here's how it typically works: An activist applies for a job (often fabricating information on the application) to gain access to a facility of interest. Once inside, this individual wears a hidden video camera. If they capture something unethical or illegal, they'll release it to the general public. People for the Ethical Treatment of Animals (PETA) used this tactic in 1981 to capture video footage of rhesus monkeys being vivisected in a Maryland medical research lab. As PETA notes on its website, the "groundbreaking investigation led to the nation's first arrest and criminal conviction of an animal experimenter for cruelty to animals, the first con-fiscation of abused animals from a laboratory, and the first U.S. Supreme Court victory for animals used in experiments."[18]

In 1992, two *ABC News* producers secretly videotaped employees for a story on improper food handling practices that accused Food Lion, a large grocery store chain, of, among other offenses, selling rat-gnawed cheese, fish dipped in bleach, rotting meat, and produce removed from fly-infested dumpsters. The airing of the television report devastated the company's bottom line: The value of its stock dropped approximately $1.3 billion in the week after the broadcast. While Food Lion denied the accu-sations, it did not pursue claims of libel or slander; rather, the company argued the two producers had illegally lied on their job applications, tres-passed, and revealed internal company information. Under this pretense, Food Lion sued *ABC News* for $2.47 billion in damages. Five years after the story aired, a jury awarded Food Lion $5.5 million, though a judge later reduced it to $315,000. Later, an appeals court threw out the entire $315,000 judgment save for $2—one dollar in damages for trespassing and one dollar for violating a duty to be loyal to Food Lion, their employer at the time.

It was in this climate that individual states began passing ag-gag laws making it illegal for activists to smuggle cameras into meat industry operations. Daisy Freund, director of farm animal welfare at the American

Society for the Prevention of Cruelty to Animals, told me that as a result, "you are lucky if you ever see past those doors if you're not personally raising animals on an industrial farm."[19] In 1990, Kansas criminalized "enter(ing) an animal facility to take pictures by photograph, video camera or by any other means" with the goal of undermining the enterprise. One year later, Montana passed an ag-gag law criminalizing "entering an animal facility with the intent to commit a prohibited act, entering an animal facility to take pictures by photograph, video camera, or other means with the intent to commit criminal defamation, and entering an animal facility if the person knows entry is forbidden."[20]

In the early 2000s, armed with huge followings on social media that made it easier than ever to disseminate videos, animal protection groups like Mercy for Animals, the Humane Society of the United States, and Animal Outlook (formerly known as Compassion Over Killing) ratcheted up their undercover investigations. To an extent it worked: Some of the worst offenders closed down and animal abusers were charged. That got industrial animal agriculture unnerved, so it attempted to pass a barrage of new ag-gag laws, with varying degrees of success. In 2002, the conservative-leaning American Legislative Exchange Council devised a piece of "model legislation" called the Animal and Ecological Terrorism Act, for distribution to lobbyists and lawmakers across the country in an effort to make ag-gag laws a widespread national phenomenon. In addition to prohibiting filming or taking pictures on livestock farms to "defame the facility or its owner,"[21] violators would be placed on a "terrorist registry." The term "terrorism" to describe such activity may seem absurd, but there is an absurd precedent for it. Brought on by the "green scare" of increased animal rights and environmental activism during the late 1990s, Congress passed the Animal Enterprise Protection Act (AEPA), boosting penalties for crimes involving the disruption of work at businesses that produce animal bodies for consumption. In 2006, Congress beefed up the AEPA by amending it to form the Animal Enterprise Terrorism Act, giving the US Department of Justice greater authority to prosecute animal rights activists.

More recently, in 2012, Missouri passed an ag-gag law that mandates evidence of animal abuse be turned over to law enforcement within twenty-four hours. While industrial animal agriculture presents this "quick-reporting" measure as a wholesome measure, it is a smokescreen designed to prevent the collection of adequate evidence to show patterns of mistreatment, neglect, or abandonment—a time-consuming process— that could help with prosecuting abusers. In 2014, Idaho passed an ag-gag law, but the Idaho District Court rejected it as unconstitutional a year later. In 2017, Texas prohibited snapping photos over factory farms with drones. And in 2019, Iowa passed a law that criminalizes using deception to gain access to an agricultural production facility with the intent to cause harm to the business. The District Court for the Southern District of Iowa issued a preliminary injunction in 2020, blocking the law's enforcement, but later that year, it passed another ag-gag law that criminalizes trespassing on a "food operation"[22] without consent. In 2021, it passed another ag-gag law that criminalizes the use of cameras or surveillance devices on farms, as well as taking unauthorized samples. There are many other ag-gag legislative efforts that have already passed or are still pending; undoubtedly more are on the way.

While checkoffs and legislative efforts that benefit industrial animal agriculture are not directly funded by tax dollars, one major program is: farm subsidies. These, in the broadest sense, are supports, including financial, that the US government gives to selected farmers and agribusinesses. The original intent was to help vulnerable farmers during the Great Depression. To help stabilize prices, in 1933, the government passed the Agricultural Adjustment Act (AAA). Additionally, the government would purchase farmers' surplus, which it could later release during any future unproductive years. Farmers who agreed to leave some of their land unplanted and ungrazed would receive a proportional amount of money from the government. To support this fund, Congress imposed a new tax on processors, such as slaughterhouses, flour mills, and cotton gins. (This was later ruled unconstitutional on the grounds that it was illegal to levy a tax on one group in order to pay it to another; a revised AAA was passed without the processing tax.) The premise was built on fundamental supply-and-demand economics: With fewer crops and livestock available to consumers, the prices for them would go back up. And that is exactly what happened: The prices of agricultural commodities rose, and by 1935, the income generated by farms was 50 percent higher than it was in 1932. Over the ensuing decades, however, a law designed to support

Depression-era farmers morphed into a multibillion-dollar apparatus supporting almost exclusively multinational corporations.

Congress continued to tinker with the AAA, adding new subsidies like crop insurance, in which the government covers losses from poor yields or declines in revenue due to natural disasters or a decline in price, essentially locking in profits. The law also provided direct payments (sometimes called "freedom to farm" payments) to farmers regardless of whether prices are high or low to theoretically encourage a stable level of production. These and other programs do not come cheap: Between 1995 and 2005, the government paid farmers an average of $16 billion per year in the form of subsidies. Although direct payments were repealed in 2014, some of the savings were added to the crop insurance program, what Vincent H. Smith, a professor of farm economics at Montana State University, referred to as "a classic bait-and-switch proposal to protect farm subsidies."[23]

While most people agree protecting farmers from bankruptcy and avoiding food shortages are worthy aims, many critics across the political spectrum debate the overall effectiveness of such subsidies. One major problem is that, like checkoff funds, subsidies are allocated disproportionally in favor of industrial animal agriculture. Demographics across time give an initial glimpse into why: When the AAA was created in 1930, nearly 25 percent of the population, roughly 30 million people, lived on nearly 6.3 million farms and ranches, according to the USDA Census of Agriculture Historical Archive. These farmers were relatively poor, with an average income of less than half of non-farmers. But, as the historian Paul Freedman explained to me, farmers in 2017 represented less than 1 percent of the population—about 3.2 million people operating 2.05 million farms. The overwhelming majority of these farms belong not to families but to industrial animal agriculture. The result is, the largest farm operations, as well as the richest farmers and landowners, with incomes many times higher than the national average, receive over 85 percent of all farm subsidies. As Robert Paarlberg states in his book *Food Politics*, "They do not need subsidies to remain prosperous, yet . . . efforts to improve the targeting of subsidy payments are routinely blocked by lobbying representing commercial farmers."[24] As one example, Paarlberg cites President George W. Bush's attempt in 2008 to prevent the delivery of some subsidy payments to farmers who earned more than $200,000: "The Senate voted that the cap should instead be set at $750,000, and the House of Representatives said there should be no cap at all."

By and large, these farms are not producing fruits and vegetables. Between 2008 and 2012, less than 1 percent of subsidies went toward vegetable and fruit growers. Rather, the subsidies heavily focus on corn, soybeans, wheat, cotton, and rice—much of which is used as animal feed. Between 1995 and 2010, about $170 billion in government money went to these commodities. In 2018, the Trump administration announced $12 billion in emergency aid to farmers caught in an escalating trade war with China, Mexico, and other countries that imposed retaliatory tariffs. The first block of aid distributed to agricultural producers was about $4.7 billion. Soybean farmers got the overwhelmingly biggest share: $3.6 billion. (The runner-up was pork, with $290 million.)

In 2019, the USDA announced another $16 billion in trade-related aid to farmers, for a grand total of $28 billon in two years. As for 2020, fueled by COVID-19, the amount more than doubled to $46.5 billion, a new record. ("Trump money is what we call it,"[25] one farmer told National Public Radio.) This, in turn, benefits the farmers, yes, but also the companies that buy commodity crops, including meat producers—what Geoff Dembicki refers to in the *New Republic* as a "de facto subsidy to the factory farming model."[26] This logic is explained by Daniel Imhoff, who writes in *Food Fight: The Citizen's Guide to the Next Food and Farm Bill* that "Tyson . . . saved nearly $300 million per year in the decade after the 1996 Farm Bill because it could buy chicken feed so cheaply. Smithfield, the world's largest hog producer, saved nearly the same amount. In total, the top four chicken companies saved more than $11 billion in the decade after the 1996 Farm Bill, while the top four hog giants saved nearly $9 billion on their feed costs."[27]

It is tempting to draw the conclusion that if it wasn't for these savings, American families would pay a lot more for meat. That was my assumption. But that isn't the case, at least not directly. Figures vary, but Purdue University's Jayson Lusk estimates that without subsidies, meat prices would be a mere 0.55 percent higher. Likewise, researchers at the American Enterprise Institute concluded in a 2017 report that the impact of subsidies "on retail prices and food consumption are surely tiny."[28]

Why? The cost of commodities like corn and soybeans that are fed to animals ultimately account for only a small share of the overall retail cost of meat. As Lusk notes, "So, while these policies may be inefficient, regressive, and promote regulatory over-reach, their effects on food prices are tiny."[29] In reality, all subsidies seem to do is fatten the wallets of industrial animal agriculture, to pay for whatever it deems necessary: high salaries, fancy marketing, lobbying, and so on, which in turn, do impact meat consumption.

Industrial animal agriculture also benefits from the Great Depression–era practice of the government purchasing farmers' surplus. When the retail price of chicken dipped in 2008, for example, the government bought $42 million worth of the industry's products, which it donated to food-providing programs and charities. The government made similar moves when it purchased $30 million and $42 million worth of chicken in 2010 and 2011, respectively; in 2018, the government announced it would shell out another $60 million; and then in 2020, an additional $30 million. Even when the demand is not there, industrial animal agriculture finds a customer in the USDA. "There's a lot of money flowing toward not just supporting the production of meat and dairy, but the overproduction of meat and dairy,"[30] Stephanie Feldstein, the Center for Biological Diversity's population and sustainability director, told *Gizmodo*. In fact, according to 2021 data from the USDA, American meat producers now have 2.1 billion pounds of chicken, turkey, pork, and beef in cold storage.

The USDA essentially has two conflicting roles: boosting the American agriculture industry and encouraging healthy eating. These two goals are most openly clashing when it comes to the USDA's Dietary Guidelines for Americans, which typically fall on the side of industrial animal agriculture. The Dietary Guidelines provide guidelines for healthy eating—MyPlate, which replaced the iconic food pyramid diagrams in 2011, is based on these recommendations—and have been published jointly every five years since 1980 by the Center for Nutrition Policy and Promotion (CNPP) in the USDA and the US Department of Health and Human Services. The latest Dietary Guidelines were released in 2020 and they tell us to fill half our plate with fruits and vegetables to maintain a healthy diet, but as we've seen, that doesn't even remotely align with how the institution actually spends its money. Moreover, while the 2020 Dietary Guidelines note "lower consumption of red and processed meats" are "common characteristics of dietary patterns associated with positive health outcomes,"[31] they don't make an explicit recommendation to eat less of it. Instead, they rely on a euphemism: "Limit foods and beverages higher in added sugars, saturated fat, and sodium," giving the food industry plenty of wiggle room to remain in school breakfasts and lunches nationwide, which draw on USDA nutritional guidelines.

Indeed, this was reflected in a 2020 proposal—one of a series of efforts by President Trump to roll back school nutrition rules implemented by his predecessor—put forth by the Food and Nutrition Service of the USDA, which, among other activities, administers the National School Breakfast and Lunch Programs. "Although cloaked in language like 'flexibility' and 'streamlining' the reality is if (realistically when) this rule passes, school food-service teams will have the green light to serve fewer fruits, vegetables, and whole grains and will have the ability to do so with even less oversight and accountability," Sanchez, the executive director of Balanced, explained to me. "There will potentially be a small number of food-service teams that take the proposed changes and shift toward healthier menus, but they will be extreme outliers."[32]

More specifically, the new rules solidified a temporary rule change quietly released in spring 2019 dictating that pasta made with starchy vegetable-based flours counted as a vegetable. It also lowered the minimum requirement of red or orange vegetables and allows schools to offer potatoes as a vegetable every day, including, yes, in the form of french fries. Schools also got permission to sell à la carte entrees more frequently. Unfortunately, the most popular à la carte options are high-calorie, high-fat, high-salt foods including pizza, burgers, hot dogs, chicken nuggets, and fries. These items also happen to be the ones that make schools the most money (because they have the highest margins). This provides a perverse incentive for schools to sell more of these options at the expense of students' health.

It certainly doesn't help that, according to Marion Nestle, the Dietary Guidelines Advisory Committee (DGAC)—the advisory committees that help shape the Dietary Guidelines—are composed of individuals who have financial ties to companies that profit from them. In her book *Unsavory Truth: How Food Companies Skew the Science of What We Eat,* she notes that historically the DGAC has written a report summarizing the most recent research and then produced dietary guidelines based on that science. But in 2005, President George W. Bush's "industry-friendly administration" changed that process: "The DGAC would continue to review and summarize the science, but the agencies would now write the actual guidelines . . . [which] took the guidelines out of the hands of scientists and put them into the hands of agency political appointees."[33] Moreover, by procuring special waivers, candidates for committee membership who consult, conduct research, or work directly for food companies, may still be eligible to serve. And ultimately, many do.

Nestle goes on to explain that although members of the DGAC must disclose conflicts of interest, the government is not required to make such disclosures public, and usually doesn't. Fortunately, we can uncover them by filing requests through the Freedom of Information Act, and reviewing members' disclosure statements in published papers. Here's what Nestle found: "In 1995, when I was on the DGAC, only three of the eleven members had ties to food companies, but the balance soon shifted: to seven of eleven members in 2000, eleven of thirteen in 2005, and nine of thirteen in 2010." More specifically, members of the 2000 DGAC reported financial ties to two meat associations, four dairy associations, five dairy companies, and one egg association, among others; in 2015, ten of the fourteen had ties to companies producing meat, dairy, or processed food; and in 2020, more than half had ties to the food industry, including two who were nominated by the National Cattlemen's Beef Association. "Overall, this looks to me like any other DGAC except that there are twice as many members as in the past,"[34] Nestle remarked on her blog.

Furthermore, as Walter Willett, professor of epidemiology and nutrition at Harvard T.H. Chan School of Public Health, explained to me, "The meat industry has funded a lot of the studies that have been done looking at red meat and, say, changes in risk factors like blood cholesterol and blood pressure, and those studies are usually designed to give a result that's favorable to the red meat industry, not surprisingly."[35] Consider a 2018 study led by nutrition professor Wayne Campbell and his colleagues at Purdue University, in which they assessed the effects of consuming different amounts of lean, unprocessed red meat in a Mediterranean diet among forty-one subjects over several months. Based on the results published in the *American Journal of Clinical Nutrition*, the researchers concluded "Adults who are overweight or obese can consume typical US intake quantities of red meat (~70 g/d) as lean and unprocessed beef and pork when adopting a Mediterranean Pattern to improve cardiometabolic disease risk factors. Our results support previous observational and experimental evidence which shows that unprocessed and/or lean red meat consumption does not increase the risk of developing cardiovascular disease or impair associated risk factors."[36]

Given the mountain of data and leading health organizations that suggest otherwise, you may find these conclusions quite surprising. That is, until you learn who partially funded it: both the Beef and Pork Checkoffs. This isn't the first time industrial animal agriculture has funded research by Campbell and his colleagues that report results in favor of meat consumption. A 2015 study partially funded by the National Pork Board concluded that hypertensive adults seeking to lower their blood pressure "can expand their protein options to include lean, unprocessed pork."[37] A year later, a study partially funded by both the Beef and Pork Checkoffs conveniently found that eating high-protein diets improves sleep in overweight and obese people. By no means are Campbell and his colleagues alone: Consider a 2015 study funded by the North American Meat Institute led by two researchers at Nutrition Impact LLC, a consulting firm with the stated goal of helping food companies "develop and communicate aggressive, science-based claims about their products and services,"[38] and one researcher in the Animal Sciences Department at North Dakota State University. It reported "that the diet quality of lunch meat consumers [was] similar to non-consumers for every age/gender group. These results suggest that lunch meats do not necessarily decrease average diet quality of adults and children."[39]

Industry funding does not necessarily mean a study's methodologies are flawed, but research nevertheless suggests that sponsored studies are much more likely to report results favorable to the interest of the sponsor in comparison to non-sponsored studies. It just makes sense: Industrial animal agriculture is unlikely to keep funding researchers who produce studies that nudge people in the direction of eating less of their product. David Katz, author of *The Truth about Food*, echoed this sentiment, explaining to me that this "pseudo-confusion" is due to a cozy relationship between Big Food and Big Pharma: "I have a morbid fantasy of CEOs from Big Food and Big Pharma smoking cigars in a boardroom with the door closed, striking a deal. Big food CEO's job is to make people sick and profit from it. Big Pharma CEO's job is to treat the unnecessary illness and profit from it. They have a good laugh and shake hands. . . . So, they're in on it."[40]

Industrial animal agriculture defends the healthfulness of its food by attempting to poke holes in sound scientific research. In 2015, for example, the International Agency for Research on Cancer, which is part of the World Health Organization, classified processed meat as a carcinogen and red meat as a probable carcinogen. Twenty-two international experts reviewed more than eight hundred studies before making this warning. Here is what Shalene McNeill, director of Human Nutrition Research at the National Cattlemen's Beef Association had to say to the *Washington Post* about the report: "We simply don't think the evidence supports any causal link between any red meat and any type of cancer."[41] The North American Meat Institute put out a statement saying the findings were a "dramatic and alarmist overreach" that "defies both common sense and numerous studies showing no correlation between meat and cancer and many more studies showing the many health benefits of balanced diets that include meat."[42]

Two years earlier, when a study led by Frank Hu, a professor of nutrition and epidemiology at the Harvard School of Public Health, found strong "evidence that red meat consumption contributes to an increased risk of diabetes,"[43] McNeill countered that there was "a lot of good scientific evidence"[44] that supports the role of lean beef in a healthful, balanced diet. Four years before that, in 2009, when a study led by researchers at the National Cancer Institute found that men and women who eat higher amounts of processed and red meat have a higher risk of overall mortality, cancer, and heart disease compared to those who eat less, McNeill sought to obfuscate the findings: "As is often the case with epidemiological research on this subject, it is hard to draw substantial conclusions about any one food,"[45] she said.

Industrial animal agriculture also deploys clever language to mitigate public concerns over the impacts of factory farms on animals and the environment. Consider the words routinely splashed across their products. "Free-range," for example, means that animals have "access to the outdoors," but the size, quality, and duration of access to that outdoor space is entirely unregulated, so conditions vary greatly. "Humanely raised," meanwhile, has not been defined by the USDA and therefore offers zero assurance about the animals' welfare. The same is true for many environmental claims. For instance, no official definition exists for "sustainably raised," another meaningless term often printed on the labels of animal products.

The same tactics extend to the role of animal agriculture in climate change. Approximately 14.5 percent of global gas emissions comes from meat and dairy, a link that the industry has worked tirelessly to blur, according to a 2021 study by researchers at New York University. "US beef and dairy companies appear to act collectively in ways similar to the fossil fuel industry, which built an extensive climate change countermovement,"[46] explained the authors of the study, published in the journal *Climatic Change*. For example, in 2009, when Congress was drafting the American Clean Energy and Security Act, which would have established a cap-and-trade system for limiting greenhouse gas emissions, Tyson Foods and other meat conglomerates worked in tandem with the fossil fuel industry to stop the bill—an effort that ultimately proved successful. Other strategies include funding research that conveniently concludes that the cattle industry's greenhouse gas emissions are negligible, or that overfishing won't deplete the world's fish stocks anytime soon.

The animal food industry has been relying on obfuscation tactics for decades. In 1970, when an article published in *Nature* postulated a link between cancer and nitrates, which are found in many processed meats, the public became alarmed. Consumer advocate and future presidential candidate Ralph Nader accused the USDA of supporting agribusiness over and at the expense of everyday Americans, citing these potentially harmful chemicals in processed meat. Nader even referred to hot dogs as "among America's deadliest missiles."[47] Leo Freedman, chief toxicologist for the Federal Drug Administration (FDA) told the *New York Times* in 1973 that although the risks involved in eating processed meats are "pretty small," he was "sure the nitrosamines are a carcinogen for humans."[48] The American Meat Institute maintained that there was no direct evidence that the chemicals in processed meats caused cancer in humans. Then, the industry came up with a redirection: It argued these added chemicals were essential because they reduced the risk of botulism, a dangerous foodborne illness—conveniently leaving out that almost all the cases of botulism in preserved food (a very rare occurrence) were the result of poorly preserved vegetables, not meat without nitrates. In 1977, the USDA and the FDA intervened, giving industrial animal agriculture three months to prove added chemicals in processed meat did not cause harm or else replace them. But under pressure from industrial animal agriculture, the FDA delayed its three-month ultimatum numerous times under the premise that more research was necessary. That research would come, and a lot of drama with it.

One day in 1978, a day the *Washington Post* referred to as "The Day Bacon Was Declared Poison,"[49] a landmark study was published by Dr. Paul M. Newberne, a specialist in animal diet and disease at Massachusetts Institute of Technology. From 1976 to 1978 he fed nitrites to over two thousand rats in his laboratory. He was evaluating, for the FDA, the proposition that nitrites cause cancer in animals. What he found seemed like something even industrial animal agriculture couldn't overcome: On average, rats fed nitrite got lymphatic cancer 10.2 percent of the time, while animals not fed nitrate got cancer only 5.4 percent of the time. On August 11, 1978, the FDA and the USDA issued a joint statement: "A study recently completed for the FDA by the Massachusetts Institute of Technology strongly suggests that nitrite produces cancer of the lymphatic system in test animals."[50] It went on to explain the two federal agencies would assess their options as they "weigh the risks associated with nitrite added to food against the health risk from not adding it [presumably referring to the bogus botulism argument]." Industrial animal agriculture cried "removal of nitrate will result in their economic ruin" while consumer activists like Nader argued, "Carol Foreman [the assistant secretary of agriculture] and Donald Kennedy [the FDA commissioner] should go back to the law and read it because they are violating it [by not taking immediate action]." After much back and forth, in August 1980, the two federal agencies cited "insufficient evidence" to say sodium nitrite can cause cancer. "I think it's a situation in which we have to wait until we know more," said Dr. Jere Goyan, the then new FDA commissioner. "I think we should eat a well-balanced, nutritious diet and not be concerned about cured meats."[51] Any hope of persuading the government fully vanished when President Reagan became president in 1981 and appointed as undersecretary of agriculture Richard Lyng, none other than the former president of the American Meat Institute; he would eventually go on to serve as secretary of agriculture. Ultimately, the proposed ban was shelved, and it has been ever since.

The "collaboration" between animal agriculture industry and governmental organizations does not end with nutritional research. As the animal agriculture industry comes under greater scrutiny for greenhouse gas emissions, it struggles to build an environmentally friendly image. I was recently reminded of the cattlemen-sponsored article I saw on *Quartz* when I encountered a study by the USDA's Agricultural Research Service (ARS), published in 2019, which found that raising cows for food is environmentally just fine: "We found that the greenhouse gas emissions in our analysis were not all that different from what other credible studies had shown and were not a significant contributor to long-term global warming," said ARS agricultural engineer Alan Rotz, who led the analysis team, in a study summary on the agency's website.[52] Needless to say, I was not surprised to discover the study was partially funded by the National Cattlemen's Beef Association.

Of course, while industrial animal ag companies benefit from a cozy relationship with government and the trade associations advocating on their behalf, they also pursue their own marketing campaigns. Enter greenwashing, the process of communicating a false impression or providing misleading information about how a company is more environmentally sound than it actually is. Consider a jointly signed, full-page ad by JBS, the world's largest processor of fresh beef and pork, and Pilgrim's Pride (largely owned by JBS), which appeared in the *New York Times* in April 2021:

Agriculture can be a part of the climate solution.
  Bacon, chicken wings and steak with net-zero emissions.
  It's possible.[53]

The companies claim they will achieve "zero by 2040" and that they have set "time-bound, science-based targets" and are "backing them up with $1 billion in capital over the next decade." Considering that JBS's annual revenues in 2019 alone were $38 billion, that the commitment is short on specifics, and that only JBS is holding JBS accountable for meeting it, I won't hold my breath, and you shouldn't either. There's a precedent for such skepticism. In September 2020, JBS announced a commitment to address its role in deforestation in the Amazon. There was only one problem: The announcement was merely a restatement of a commitment made in 2009, a pledge it failed to deliver on in the subsequent decade. "The industrial beef sector is a liability," said Daniela Montalto, forests campaigner with Greenpeace in response. "While promising to maybe someday save the Amazon, JBS and the other leading beef processors seem willing to butcher the Pantanal [the world's largest tropical wetland] today, making mincemeat of their sustainability pledges."[54] So it's no surprise that environmental advocates like Sarah Lake, vice president and global director for Latin America at Mighty Earth, is skeptical about this new 2040 commitment: "JBS has repeatedly made false promises."[55] So, no, this, I'm afraid, is likely nothing but a weak attempt at making the company seem eco-conscious, when it is anything but.

Industrial animal agriculture's promotional strategies, often in the face of inconvenient research, are vast; its *modus operandi* is entrenched in the very foundations of our political, legal, and educational systems. From savvy marketing and public relations campaigns aimed at the general public to strategic lobbying efforts and partnerships, the industry not only creates demand, but more subtly, removes any potential roadblocks to filling that demand, stacking the odds in its favor well before any of its products hit the shelves. The takeaway is clear: Industrial animal agriculture does all it can to keep us hooked on meat.

9

The Meat Paradox

I don't eat red meat, but sometimes a man needs a steak. —*Gwyneth Paltrow*

Remember when the Biden Administration banned meat?

The media frenzy began when the *Daily Mail* reported breathlessly that "Biden's climate plan could limit you to eat just one burger a MONTH."[1] Florida Rep. Marjorie Taylor Greene immediately dubbed President Biden "The "Hamburglar,"[2] as the far right-wing media world unloaded on the White House's apparent plan to end meat as we know it. Fox News host Jesse Watters claimed that "President Biden has been boasting about his plan to save the planet and cut carbon emissions by 50 percent. . . . Americans would have to cut red meat consumption by a whopping 90 percent. That means only one burger a month."[3] Donald Trump Jr. proudly tweeted, "I'm pretty sure I ate 4 pounds of meat yesterday. That's going to be a hard NO from me."[4] Colorado Rep. Lauren Boebert demanded that Biden "stay out of my kitchen."[5] Tweeted Texas Governor Greg Abbott: "Not gonna happen in Texas!"[6]

There was only one problem: The Biden administration never proposed any limit on Americans' meat consumption. The *Daily Mail*, which had initially sparked the Republican outrage, baselessly connected Biden's proposal to an academic paper from 2020 that had nothing to do with

Biden or government meat bans. According to CNN, scholars at the University of Michigan and Tulane University estimated "how greenhouse gas emissions would be affected if Americans hypothetically decided to change their diets in various ways, such as cutting their consumption of beef to four pounds per year."[7]

There's a lot we can take away from this episode, notably that Americans—particularly far-right conservative Americans—*really* like their hamburgers. It's reminiscent of when the Hollywood actor–turned–National Rifle Association president Charlton Heston dared the government to pry his gun "from my cold, dead hands." Biden wasn't just coming for America's guns; he was coming for its hamburgers, too.

Is meat just one more fault line in a country that is ripping itself apart over hotbed issues like abortion, guns, cancel culture, vaccinations, and immigration? There is some evidence for this. In early 2021, progressive icon Alexandria Ocasio-Cortez asked her twelve-million-plus Twitter followers to join her in going vegetarian for Lent. Months later, Vice President Kamala Harris, at the urging of her friend (and vegan) Senator Corey Booker, visited a vegan taco shop and revealed she is trying the "Vegan Before 6 p.m." plan. And in July 2021, Brooklyn Borough President Eric Adams, an avowed vegan, won the Democratic nomination for mayor of New York City. According to data from Pew, Democrats are about twice as likely to be vegetarian or vegan compared to Republicans. One psychological study speculated that right-wingers consume meat for two main reasons: "(a) they push back against the threat that vegetarianism and veganism supposedly pose to traditions and cultural practice, and (b) they feel more entitled to consume animals given human 'superiority.'"[8] In other words, do conservatives eat more meat simply because they are sticking it to liberals and animals?

Texas Senator Ted Cruz once posted a video of himself cooking bacon wrapped around a gun barrel. Conservative pundit Matt Couch tweeted a picture of his dinner plate stacked with four pounds of meat (and, for good measure, asparagus wrapped in meat). Jordan Peterson has been publicizing his daughter's beef-and-water diet and $90-per-hour carnivore consulting sessions. Alex Jones in 2016 tweeted a picture of him salivating over a plate of sausages and steaks with the caption: "Celebrating Americana with some Red Meat, f-you Obama!"[9] President Obama was neither a vegetarian nor had he advocated for eating less meat, but as the *Daily Beast* pointed out, "The implication was clear: Obama and his ilk wanted to take away America's meat, just like they were coming for guns and Confederate flags and all manner of totems beloved by conservatives who believe themselves besieged by the left."[10]

This political schism is not lost on meat producers. According to OpenSecrets.org, 79 percent of all contributions to federal candidates by the meat processing and products industry have gone to Republicans since the 1990 election cycle.[11] In 2020, the livestock industry donated $14.1 million to Republicans, compared to just $4 million to Democrats, with Donald Trump receiving nearly $2 million to Joe Biden's $413,000. The same trends hold with nearly every sector of the agriculture industry, from dairy to poultry to eggs. And yet these are paltry sums when you consider that the meat industry has annual sales in the United States higher than the gross domestic product of Hungary and Ukraine. Interestingly, despite the industry's overwhelming preference for the Republican Party, the Biden administration "has steered clear of discussing stricter environmental regulations that could scare off the largely conservative farm sector, as well as the rural lawmakers that Biden will need to advance many of his environmental goals,"[12] *Politico* reported in April 2021, despite the farm industry representing a massive 10 percent of all US emissions.

However, while liberals are certainly more likely to be vegetarians or vegans compared to conservatives, the vast majority of left-wingers are just as carnivorous as their counterparts on the right. According to Gallup, 11 percent of liberals identify as vegetarian while just 5 percent are vegan. For all intents and purposes, eating meat is a bipartisan affair, and the psychology of America's meat obsession runs a lot deeper than red plates and blue plates.

Back in 2015, the University of Lancaster attempted to explore the psychology of meat consumption, particularly how people are able to snuggle up with animals like dogs, cats, horses, and baby pigs—and furthermore, empathize when they are not treated well—and then go to a restaurant and eat steak, pork, and chicken. According to the study, these justifications fall into four main categories, which researchers call the "Four Ns." As the *Sydney Morning Herald* explained, "Meat eating is Natural ('humans are natural carnivores'), Necessary ('meat provides essential nutrients'), Normal ('I was raised eating meat') and Nice ('it's delicious'). Of all these reasons 'necessary' was the most common."[13] Interestingly enough, the people most likely to endorse all four Ns were much more likely to be men, who "also felt that cows were less likely to experience feelings like sadness and joy," explained one of the researchers.

If political preference doesn't fully explain the differences between carnivore and vegetarians, could gender? Could that explain why Jesse Waters chose to eat a steak while debating toxic masculinity with a feminist vegan on his Fox show?

If you search for "preparing salad" on a stock image site, most of the images will feature women. But when you type "grilling meat," you get burly men flipping burgers in their backyard. As Marta Zaraska, author of *Meathooked*, told the *Daily Beast*, "The connection between meat and masculinity has been with us for 2.5 million years, basically since we've started scavenging for rotten zebras on the savanna. . . . For centuries men limited women's access to meat—in many cultures there are taboos on women eating certain types of meats, sometimes even punished by death."[14]

After I graduated college in 2011, I signed up for the dating app OkCupid and began a long string of awkward first dates—forty-five over the course of a year, in fact, before I finally met my future wife, Isabel. One of those painful first dates had a lasting impact on me, however. We were at a restaurant making small talk when the waiter came to take our orders. My date, Rachel, ordered a salad; I ordered a medium-rare steak.

She stared at me. "I thought you were a vegetarian," she said, almost accusingly. Indeed I had stated as much in my OkCupid bio, though at this point in my life I was more of an aspirational vegetarian. I sputtered and tried to explain that the app didn't give an option for "flexitarian," and that I only ate meat on special occasions. ("Like this one!" I said, lamely.) The truth was, I had ordered meat in many of my not-so-special first dates, even when there were many delicious plant-based options on the menu.

Finally, I admitted the truth of it: "I think I ordered meat because I wanted to seem manly." My voice may even have squeaked a bit.

Rachel smirked and continued to press the point. She asked what was "manly" about eating a steak cut from a powerless cow who had spent the final months of its life confined in a feedlot before being butchered in a slaughterhouse. What did eating factory-farmed flesh, or any flesh for that matter, have to do with manliness? Was eating plants somehow feminine, she asked, and by extension "weak?"

I felt myself sinking into my chair as she continued her broadside. I had no good answers. After this first date—which, needless to say, was our last—I took a hard look at the connection between meat and gender roles. How did this construct that meat is masculine and plants are feminine emerge? I certainly wasn't alone in buying into this fallacy. For instance, the musician James Blunt, known for his tender vocals, recently confessed in a podcast that in college he was so insecure in his manliness that he ate solely chicken, beef, and mayonnaise for eight weeks until he developed scurvy due to a lack of vitamin C.

In one experiment led by Paul Rozin, a professor of psychology at the University of Pennsylvania, researchers asked students which foods were the most "male" and "female." The students ranked steaks, hamburgers, and beef chili as the most masculine, with peaches and chocolate as the most feminine. Rozin and his colleagues concluded based on these and other experiments that there is "consistent, converging evidence that mammal muscle meat stands in some positive relation to maleness."[15]

But what does "manly" mean, exactly? If you watched those classic George Foreman commercials featuring the former heavyweight champ and his Lean Mean Fat-Reducing Grilling Machine, you might think that you have to eat lots of protein to be strong and burly. Every plant-based eater has been asked by their concerned friends the classic question, "But where do you get your protein?" This stems from the pervasive myth that meat and dairy products are the sole sources of dietary protein—and therefore the only way to build and maintain muscles. (Any doubts about the ability to excel in athletics while on a plant-based diet should be put to rest by the likes of MMA champ James Wilks, former NFL defensive lineman David Carter, pro cyclist and Olympian Dotsie Bausch, ultrama-rathoner Scott Jurek, tennis legend Venus Williams, US national soccer team co-captain Alex Morgan, and NBA superstar Kyrie Irving.) Even the language we use every day is coded in these meat-strong, plants-weak terms. When you go to the gym and work out, you're "beefing up" your muscles. If you stay home on the couch and watch Netflix, you're being a "couch potato." A fine physical specimen is a "beefcake." A literally lifeless person is a "vegetable."

Growing up in Staten Island, New York, I attended a lot of barbeques. Memorial Day, July 4th, Labor Day—every holiday was an excuse to fire up the grill and invite over friends. If we were grilling at my house, it was my dad on the grill. If I went to a friend's house, it was their dad on the grill. No matter where I was, the men congregated around the grill and the women made brownies, prepared the toppings, or mixed drinks. This stereotype is endlessly reinforced by TV ads. Take, for example, a 2009 spot by the charcoal supplier Kingsford, which begins with a woman pouring charcoal into a grill. When she is spotted by her husband, he quickly stops her.

"Honey, this isn't a stove. What would happen if I just walked into the kitchen and started making a salad?"[16]

"That would be weird," his wife concedes.

In 2010, Meghan Casserly, then a staff writer for *Forbes* magazine, tried to deconstruct the gender stereotypes surrounding grilling. On one hand, "grilling is exciting," she wrote. "You've got lighter fluid, a match, a breeze and a miniature pitchfork to stab things with. The potential for danger is large—and thrilling." Second, grilling was synonymous with hanging out with the guys. "The grill provides entertainment for the men, who tend to congregate around it, usually with beers. Almost like a flat-screen television during football season."[17] But as Casserly further explains, grilling is a uniquely twentieth-century American experience. In other countries, the simple act of cooking a hunk of meat over an open fire remains women's work. It wasn't until (predominately white) Americans left for the suburbs throughout the 1950s and 1960s that backyard barbeques became popular—and with it, the trope of the hardworking male who mans the grill on weekends to spend quality time with the family.

Perhaps the landmark text on the relationship between gender and meat-eating is Carol J. Adams's *The Sexual Politics of Meat*. Published in 1990, the book argues that the food we eat—namely, animals—is determined largely by patriarchal politics. Adams explains that behind every meal is an absence: the death of the animal whose flesh we are now consuming. This "absent referent" functions to "cloak the violence inherent to meat eating," and to keep our meat "separated from any idea that she or he was once an animal, to keep something from being seen as having been someone."[18] Likewise, Adams argues that society similarly reduces women to their body parts while disassociating with their humanity. Wings, thighs, and breasts—are we discussing menu choices or a human who is valued solely for her body parts? Thus, the theory goes, a patriarchal society obsessed with treating women as flesh is similarly obsessed with eating flesh.

Part and parcel with the relationship between gender and meat is the influence of class. For as long as humans have been picking at the carcasses of other animals, we have been sorting ourselves into the haves and have nots. As discussed earlier in the book, for nearly all of human history, meat was a rare luxury—and, therefore, highly sought after. This involves a psychological principle called scarcity, in which humans place a higher value on something—in this case, meat—that is scarce, and a lower value on objects that are in abundance. In basic economics this is also known as supply and demand, but the same psychology applies to early humans fighting over a half-eaten gazelle carcass or teenagers lusting after a PlayStation 5.

Take the diet of a typical European farmworker in the 1700s: bread, cheese, butter, tea, milk, and tough root vegetables. Meat was seldom, if ever, on the dinner plate. Compare that with the aristocracy, who devoured up to three pounds of meat and fish every single day. In nearly every society until the twentieth century, the food a person ate was symbolic of their socioeconomic class, turning meat into an aspirational object. As discussed earlier in the book, at least in the United States, meat is abundant and cheap thanks to industrial farming. While today's status symbols have wheels and bling, studies show that people with lower social status still tend to desire meat more than people who are better off. In one experiment led by researchers Dr. Eugene Chan and Dr. Natalina Zlatevska, subjects were shown either a meat or vegetable patty and asked which they preferred. In every case, participants who were less affluent demonstrated a much higher desire for the meat-based burger. Dr. Zlatevska surmised that class norms influenced these findings: "There is a symbolic association between eating meat and strength, power and masculinity. It is traditionally a high-status food, brought out for guests or as the centerpiece of festive occasions."[19]

Dr. Zlatevska's remark about "festive occasions" brings up another critical factor when it comes to the psychology of meat consumption: religion.

When I was growing up, Judaism generally played little role in dictating my family's behavior. We didn't follow many of the 613 *mitzvahs* ("commandments" in Hebrew). We didn't "honor God" by covering our heads with yarmulkes such that "the fear of heaven may be upon" us (Ethics of the Fathers 1:3). We didn't observe *Shabbat*, a weekly twenty-five-hour observance from just before sundown each Friday through nightfall Saturday commemorating God's creation of the universe, on the seventh day of which God rested from his work. We almost never went to Temple—even on *Yom Kippur* ("The Day of Atonement" in Hebrew)—the holiest day of the year. And we certainly didn't keep kosher ("fit" in Hebrew, as in "fit for consumption") by abstaining from pork or shellfish; bacon and cheeseburgers were staple foods growing up.

We did, however, celebrate the festive religious holidays where food and family played central roles. On Rosh Hashanah ("The Jewish New Year" in Hebrew) we would dip apple slices into honey to symbolize our hopes for a "Sweet New Year." During Hanukkah ("dedication" in Hebrew) we would eat potato latkes fried in oil to honor the miracle of one night's oil lasting eight nights in the Holy Temple in Jerusalem. Still, for my family, no religious observance exemplified the link between food and Judaism more than Passover.

Every year we'd celebrate at my uncle and aunt's home in New Jersey. My uncle began the proceedings by passing around a basket filled with yarmulkes. A Haggadah ("telling" in Hebrew)—an instructional guide for conducting the Seder—was on top of each of our plates. We each participated in the rituals as my uncle passed around the foods for us to eat, like *maror* and *chazeret* (herbs symbolizing the bitterness and harshness of the slavery the Hebrews endured in Egypt) after dipping them into *charoset*, a sweet, sticky paste of nuts and fruits representing the mortar and brick used by the Hebrew slaves to build the storehouses and pyramids of Egypt. The only meat on the Seder plate was a piece of roasted lamb or chicken. In the story of Exodus, God inflicted ten plagues upon the ancient Egyptians. To spare the Jews from the final plague (the death of the Egyptian first-born), they were instructed to mark their doors with the blood of a spring lamb; God would then pass over the first-born in these homes. Hence, the meat symbolizes the memory of the sacrificial lamb; it is not eaten or handled during the Seder. My aunt would then proceed to serve traditional kosher Jewish food, including gefilte fish, matzo ball soup, roast chicken, potato kugel, and tzimmes (a stew of carrots, raisins, and sweet potatoes). As a picky eater, I rarely ate the Seder food; instead, I would usually eat two slices of pepperoni pizza in the car that I convinced my parents to buy for me on the way. Nevertheless, I grew up with the understanding that meat was a very special luxury that was deeply important to my family's faith.

During my very short-lived tenure at Hebrew School, I learned about Noah's ark and somehow made the connection between chickens and the drumstick I was eating. I asked my dad, "Does God want us to eat meat?" Immediately after dinner, we went to Blockbuster Video and rented an animated movie, "The Beginners Bible: The Story of Creation." I already knew about the Creation Myth, but after watching it and then discussing it with him, I realized my dad was eager to highlight a subtler point that had escaped me: God not only created the universe and everything within it, but he also did so *for* humans, he explained.

"Even the animals?" I asked.

"Yes, every living thing was made here for man."

"Why?"

"He could have made it so all we have to do is take a pill, but he wanted everything to be interrelated," Dad explained.

"What do you mean, *interrelated*?"

"Somehow there is a spiritual component to food—you elevate things when you eat it. When you eat a cow, you're elevating that cow."

"That doesn't make sense," I retorted.

"God said it, so it's the end of the discussion," he said calmly but with authority. "We don't understand everything he says."

This attitude certainly isn't limited to the Jewish faith. Christians traditionally eat lamb on Easter and ham on Christmas. Eid Al-Adha ("Festival of the Sacrifice" in Arabic) is celebrated on the tenth day of Dhu al-Hijjah ("The Month of the Pilgrimage" in Arabic) and lasts for four days, during which Muslims usually sacrifice a sheep or lamb and distribute its meat among family, friends, and the poor. The sacrifice is meant to commemorate Abraham's willingness to sacrifice his son as an act of obedience to

God's command—a willingness that God acknowledged before stopping Abraham and provided him a ram to sacrifice instead.

Mankind's dominance over animals is quite explicitly enshrined in the Bible. After God created the heavens and the earth, he made Adam from dust and allowed him to enter the Garden of Eden, where inhabitants only ate plants and were free of violence. "I give you all plants that bear seed everywhere on the earth, and every tree bearing fruit which yields seed: they shall be yours for food," God said. "All green plants I give for food to the wild animals, to all the birds of heaven, and to all reptiles on earth, every living creature." When God decided to return Earth to its pre-creation state of watery chaos because it had become corrupt and violent, he instructed the only righteous man left, Noah, to build an ark and preserve two of every animal from the world-engulfing flood. Afterward, God blessed Noah and his sons, telling them:

> Be fruitful and increase in number and fill the earth. The fear and dread of you will fall on all the beasts of the earth, and on all the birds in the sky, on every creature that moves along the ground, and on all the fish in the sea; they are given into your hands. Everything that lives and moves about will be food for you. Just as I gave you the green plants, I now give you everything. (Gen 9:3–4)

Not only did God permit humans to eat animals; he also included a scroll-full of stipulations. He detailed which animals are acceptable for people to consume, including those who have cloven hooves and chew their cud (a portion of food that returns from a ruminant's stomach to the mouth to be chewed for the second time), such as cattle, sheep, goats, and deer; chicken, turkey, quail, Cornish hens, doves, geese, duck, and pheasant; and seafood as long as it has fins and scales. Presumably to avoid any doubt, God also detailed which animals were not acceptable to consume. These included pigs, camels, rabbits, lobsters, oysters, shrimp, octopus, clams, and crabs, all reptiles and amphibians and most insects (except certain locusts), as well as certain parts of all animals, including the sciatic nerve in the hindquarters and blood. He also insisted certain procedures (known as *shechita*) be followed when slaughtering an animal; only then could the meat be considered kosher. In particular, a trained ritual slaughterer (known as a *shochet*) must use a single slice of an extremely sharp knife that has no nicks, dents, or imperfections so the animal's throat is cut while they are still alive, and all their blood is drained.

Of course, not all of the world's religions are so intertwined with the consumption of meat. Seventh-Day Adventism is a denomination of Christianity that strives to mimic the original vegetarian diet practiced in the Garden of Eden. In contrast to Peter's dream, which overturned the post-Noah era set of food rules for culinary libertinism, Seventh-Day Adventism founder Ellen White had a vision in which God set back the clock to the more restrictive pre-Noah standards, but with a nod to post-Noah, pre-Christian sensibilities: God's people were to abstain from eating all animal flesh, but especially pig flesh. So many Adventists are strictly vegetarian from birth that they are often the subject of scientific studies on the long-term health effects of plant-based eating. On the other side of the globe, religions originating in East, South, and Southeast Asia have their own take on meat consumption. The Vedas—the oldest scriptures of Hinduism—recommend ahimsa ("not to injure" in Sanskrit)—the concept of nonviolence against all life forms including animals. In the ancient Indian religion Jainism, vegetarianism is mandatory, but the restrictions go even deeper than avoiding flesh. Many Jains not only avoid eggs and honey, but subterranean vegetables too, viewing the bulbs of the underground plants as living creatures due to their ability to sprout.

Nevertheless, for billions of people around the world, religion has elevated the status of meat into a special food eaten during celebrations, whether it's Easter brunch, Seder dinner, or Eid Al-Adha. This has the effect of turning meat into a reward that we crave. Marta Zaraska points to experiments in which children were told they could only have certain snack foods if they behaved well. "Right away they started craving these snacks more than did kids in a control condition, who were just simply offered the food," she explained. Research out of the University of Georgia has shown that if you give kids snacks they don't necessarily like, but do it while being extremely friendly, children will actually start to enjoy them. Considering that our biggest holidays, like Christmas, Thanksgiving, and Independence Day, are heavily associated with eating meat, it may be no surprise that we have come to enjoy and crave them.

While political preferences, gender norms, class, and religion certainly play a major role in preferences for meat, there may also be a much simpler explanation for our meat addiction, one rooted less in psychology and more in genetics. A few years ago, researchers out of Cornell University wondered why some people just seem to take to plant-based diets well, while others struggle mightily. Scientists discovered a genetic variation that allows people to "efficiently process omega-3 and omega-6 fatty acids and convert them into compounds essential for early brain development."[20] This variation, called the vegetarian allele, is widespread among populations like India that for millennia have subsisted primarily off fruits, vegetables, and whole grains with very little meat. In total, 70 percent of South Asians, 53 percent of Africans, 29 percent of East Asians, and just 17 percent of Europeans had the vegetarian allele. In the United States, just 18 percent of those studied had the vegetarian allele, likely because many Americans have Northern European ancestry, and are more likely to eat meat and have "a long history of drinking milk," said the researchers. In other words, some populations have been eating meat and dairy for so long that their bodies have adapted. Conversely, those who have primarily subsisted on plants have bodies that better process omega-3 and omega-6 fatty acids, which are found in plant foods like nuts, seeds, and avocadoes.

All told, there is no simple answer to the "meat paradox"—that is, why people continue to eat meat in large quantities despite knowing that it is neither healthy, humane, nor environmentally sustainable. Whether it's due to their political ideology, perceived expectations about their gender, socioeconomic status, religious beliefs, or their DNA, most people eat meat every single day. Moreover, it's an incredibly difficult habit to kick. I've learned this the hard way over the years. In 2021, the Reducetarian Lab, the research arm of the Reducetarian Foundation, in collaboration with several colleagues, published a study in the *Journal of Environmental Psychology* in which we tried to gauge which strategy was more effective: convincing people to give up meat entirely or to cut back. In the pilot, we asked several thousand people to read an op-ed advocating for either a complete elimination of meat from their diet or a modest decrease in their meat consumption. When we checked back five months later, the participants who were asked to eliminate meat categorically reported no significant change in their eating habits. Those who were asked to scale back their consumption, however, on the whole were consuming 7 to 9 percent less meat five months later. But when the study was scaled to a nationally representative sample, the effect disappeared.*

Our study and other ones like it point to a simple conclusion: It's not easy giving up meat, and we don't like being told to do it all at once. Perhaps, as we'll see in part 3 of this book, the key to a sustainable, healthy, and humane future isn't eliminating meat, but reimagining it.

---

* It wasn't until we did a deeper dive into the data that we notice a pattern: Effects were found only for younger, liberal, more educated, less wealthy participants—demographics similar to those studied in the pilot.

III

MEAT OF THE FUTURE

10

Turning Back the Clock

To cherish what remains of the Earth and to foster its renewal is our only legitimate hope of survival. — *Wendell Berry*

I'm on my way to Bluffton, a small town of just over one hundred residents in the deep south of Georgia. I'm meeting with a legend among foodies: a fourth-generation farmer named Will Harris. Will owns White Oak Pastures, a sixth-generation, century-and-a-half-year-old 1,250-acre family farm considered to be one of the most environmentally friendly and humane farms in the country, maybe even in the world, and a stark contrast to the factory farms dominating today. I had reached out to Will a few weeks before, and he kindly agreed to give me a tour.

I awake before the crack of dawn and head out to meet Will at his office. When I arrive, a man steps out of his pickup truck into the hazy morning light. "Are you Brian?" he asks in a Southern drawl, extending his hand. "I'm Will."[1]

Once we enter his office, I get my first real look at him: a sturdy, late-middle-aged man with a salt-and-pepper mustache and matching goatee, rectangular-framed glasses, a white cowboy hat, blue jeans, brown boots, a tan long-sleeved shirt, and a forest green vest. A hodgepodge of strange artifacts covers the surface of nearly every piece of wooden furniture: everything from obscure books and black-and-white photos to a dozen bottles of liquor and four or five shotguns. Nervously I mention that I have never seen so many guns in my life, to which he replies with a playful smirk: "You know what they say about guns, right? They are like golf clubs—you can never have too many."

It dawns on me that while I'm here to tour his award-winning farm, Will is clearly the main attraction. He is clearly not a person to be messed with, yet his energy and charm are palpable. He's a natural-born leader and rugged schmoozer. He invites me to take a seat next to him at a long table in the corner of the room. This is when the dynamic takes a bit of a turn. Will explains bluntly that while the tour is still on, he has reservations about me being here. He did his homework and knew I had spent the past few years advocating that people cut back on animal products. I do my best to assure him that I know his farm is not a factory farm and that I'll give him and his operation a fair shake.

Seemingly satisfied with my response, he gives me some background on his family and farm. Along the way, he explains the origins of some nearby items; Will has a knack for storytelling and clearly this is not his first rodeo. For example, while holding a Union Artillery Officer's sword that somehow came into the possession of his great grandfather James Everett Harris, a Confederate cavalry captain in the Civil War, Will tells me with pride that the farm had been in his family since James came to Bluffton in 1866. James's grandson, Will Bell Harris, Will's father, turned the farm into an industrial cattle farm after World War II, when traditional methods of farming began to disappear and "commoditization, industrialization, and centralization shot up." Then, about two decades ago, despite the fact that he "was very successful financially raising cattle industrially," Will gradually transformed the farm back into one his great-grandfather would recognize. He did this in service of "animal welfare, regenerative land management, and rural revival of this little community . . . [which] all fit together; you can't build the two without the other." While he used to think only about pounds of animal flesh and how cheaply he could produce them, that thought no longer crosses his mind; instead, he only thinks about farming strategies that improve what he calls "a complex living unit."

To ground his point, he shows me a book by a dear friend of his, *Dirt to Soil: One Family's Journey into Regenerative Agriculture* by Gabe Brown, and explains to me the principles of regenerative agriculture, what he describes as a "holistic approach to farming." One of the primary functions of regenerative agriculture, Will says, is to improve soil health, primarily through a variety of farming practices that increase organic matter in the soil. One of these key practices is to use rotational grazing—the practice of moving animals between pastures on a regular basis—to mimic the natural herds of large grazing animals that once lived abundantly in the wild. (Overgrazing weakens root systems and eventually causes plants to die, whereas rotational grazing allows plant life to rest and recover.) This is important, Will notes, because their waste provides critical nourishment for soil microbes that stimulate overall productivity. He calls this "animal impact," a tool for regenerating the land.

Among other environmental benefits, Will says regenerative agriculture not only increases animal and plant diversity in the soil, but also boosts the soil's water-holding capacity while sequestering carbon at greater depths, helping mitigate climate change. He shows me some preliminary data from an independent analysis of his farm as proof. Then he shows me two jars of soil: One is deep black, finely textured, and rich in nutrients and moisture. The other jar is depressingly light brown, hard as a rock, and lacking any semblance of life. He explains the superior one was created using animal impact while the other one was not. The deep black soil was alluring, and I was excited to see it all in action.

As I would later read, some scientists do agree with Will's positive take on regenerative animal agriculture. Though it's certainly not the case that all grass-fed cows are involved in regenerative practices, cows who eat grass for their entire lives are actually helping to increase carbon sequestration. As they graze, they cause grass to grow deeper roots, which enables it to capture more carbon than it otherwise would. But these cows have an environmental cost as well: They burp methane into the atmosphere. Methane is a shorter-lived greenhouse gas than carbon, but it's also much more potent. Because grass-fed cows grow more slowly than their factory-farmed counterparts, they live longer lives and thus expel more methane into the atmosphere. Given all the tradeoffs, it's unclear whether grass-fed cows have a net positive or negative impact when it comes to combating climate change.

We leave his office, hop into his pickup truck (which I notice has several loose handguns in the backseat), and make our way to the farm as warm sunlight spills over the horizon. As we drive through town, Will explains that Bluffton is one of the poorest counties in the nation. He blames the centralization of industrialized animal agriculture for ruining what was once a thriving agrarian economy. Like some sort of bucolic Rockefeller, Will now effectively owns nearly everything in Bluffton "except the post-office." As we approach his farm, I am instantly struck by its beauty and vastness. It's picturesque in every sense of the word: Beneath endless stretches of blue sky is the most emerald grass imaginable, interspersed with swathes of rich black soil.

Over the next few hours we move from one part of the farm to another, and I meet some of the happiest animals I've ever seen: pigs gleefully splashing around in the mud, chickens pecking around freely socializing with one another, cows chewing their cud as they gaze upon the open pasture. Will tells me that unlike animals on factory farms, these animals have genes from heritage lines, meaning they don't grow abnormally fast or large. They are pasture-raised completely outdoors and roam freely throughout their entire lives. In this way, they are able to "express their instinctive behavior." Unlike factory-farmed cattle who are fed corn in a feedlot for about five months before they are slaughtered, these cattle eat exclusively grass. The chickens dine on bugs and grubs from the pasture as well as corn, soy, barley, and other grains, while the hogs enjoy cracked eggs from the pastured egg operation and vegetables from an organic vegetable farm (produced without using synthetic pesticides or fertilizers). White Oak Pastures is also certified by Animal Welfare Approved, Certified Humane, and Global Animal Partnership—third-party verification programs that exist along a broad spectrum but can indicate the animals lived significantly better lives than on industrial farms. And I can see it with my own eyes: These farmed animals have won the jackpot.

Will introduces me to various friendly staff members, most of whom were born in Georgia. Surprisingly, however, there are also a couple of recent Ivy League graduates from distant states who came to work on the farm because they were drawn to its mission. Will tells me he's proud to be the largest private employer in the county, with 165 employees who make about twice the county average. All in all, it seems like one of the happiest places on earth.

I expected my impression would drastically change, however, during the part of the tour I had been dreading: a visit to the farm's slaughter-house to observe the killing of a cow. Will tells me the slaughterhouse I am about to see was created in consultation with Temple Grandin, the renowned autism rights activist and animal scientist who teaches at Colo-rado State University and works to improve animal welfare in farming. Grandin designed the slaughterhouse to maximize animal welfare and minimize animal stress. As we enter, I see blood everywhere. As it dawns on me what I am about to witness, I struggle not to gag. I had earlier asked Will what he thinks of the slaughter, and he explained that while this is the most humane way to take a life, "anybody that enjoys watching slaughter or watching anything die has something wrong with them—it's not a spectator sport."

After waiting a few minutes, Will warns me the grim moment is about to begin. I ask him to narrate as it happens. I hear clattering and see a cow with her eyes wide open. I hear a loud click—the sound of a bolt piercing the cow's skull and brain to render her insensible. The cow collapses and is then hoisted into the air by a machine. An employee takes a large knife and cuts her underbelly open. Blood pours onto the ground. The innards are removed, the hooves are chopped off and tossed into a bin, and then the cow's skin is cut off. (It will later be transformed into leather.) Even-tually, the cow's carcass will be broken down into specific cuts of meat. Although I am horrified by every moment of the slaughter, I take bitter-sweet refuge in knowing that up until this moment, unlike on factory farms, the cow had a very happy life.

Meat from White Oak Pastures is considered the gold standard of what is often referred to as "better meat." The meaning of this phrase varies widely depending on the person you ask, but proponents generally con-tend that in comparison to factory-farmed meat, better meat is far more sustainable (or even regenerative), and is more compassionate, creates more jobs, is healthier and safer to consume, and is even tastier. But can better meat actually compete with factory-farmed meat?

Many critics of better meat argue that it is so expensive that only the wealthy can enjoy it. The precise cost of better meat varies widely depending on production practices and the quality of cut, but let's consider some better forms of beef. According to USDA data, in March 2021, the average per pound retail cost of grass-fed ribeye steak was $22.61. Traditional beef? $8.26. No matter the cut, grass-fed beef is about two to three times more expensive than the commodity variety. As another example, as of writing, a "grassfed beef flank" ten-ounce steak from White Oak Pastures is $14.99; a "Tyson Angus Choice Flank" ten-ounce steak from Walmart, is $5.20—nearly one-third the price. The reason better beef is pricier has to do with beef producers' profit margin: It takes up to a year longer for a grass-fed animal to reach slaughter weight compared to an industrially raised one—not to mention much more food, care, and labor. Despite that extra time, grass-fed cattle are usually smaller than grain-fed cattle at slaughter, which means less meat to sell per head. Not surprisingly, other meats follow a similar price differential.

Proponents of better meat say despite the increased cost, there is clearly a niche market for it: According to Nielsen data, retail sales of fresh grass-fed beef increased from $17 million in 2012 to $272 million in 2016. (A separate report from SPINS found grass-fed meat was up to $254 million in 2019 from 181 million in 2017.) But even when considering the overall grass-fed market—including fresh retail, packaged foods, and food service—it accounts for only an estimated $1 billion, less than 1 percent of the overall $105 billion total US beef market. (Regenerative-branded meat is too new for reliable market data, but it's likely considerably less than grass-fed.) Will himself acknowledges that in the end, it's consumers who must decide whether they "care enough about the land, animals, community and environment"[2] to pay more for meat produced using better practices. (The only other option to cover the costs, as Daniel Sumner, professor of agricultural economics at the University of California at Davis explains, is "some collective action or government program,"[3] such as subsidies or regulations, that helps farmers sell their goods at a less expensive price, effectively putting all farmers on an equal playing ground.)

Another reason many skeptics call into question the scalability of better meat is that even if price parity could be achieved, there isn't enough land to support enough production of it at current consumption levels. "It turns out that there is no easy way to replace all of the feedlot beef production in the United States with grazing cattle,"[4] writes Jonathan Foley, executive director of Project Drawdown, who goes on to cite a study by researchers Matthew Hayek and Rachael Garrett showing "current pastureland grass resource can support only 27% of the current beef supply."[5] However, many advocates of better meat—especially within regenerative agricultural circles—say this simply isn't true: As one example, according to a 2017 report produced by Stone Barns Center, a nonprofit farm and educational center, an analysis of US land resources indicates that "if the 17.6 million acres currently used to grow grain for feedlots were converted to pasture, there would be enough land to [raise] 33 million head of cattle, very close to the targeted 36.6 million,"[6] that is, the amount of cattle needed to produce the equivalent amount of beef currently consumed from grain-fed cattle per year. To make up the difference, the authors point to other strategies, including the grazing of idle grasslands as well as some land currently enrolled in the USDA Conservation Reserve Program.

In response, many skeptics of better meat argue that even if it's *theoretically* possible to scale grass-fed beef production in the United States, there are simply too many obstacles to overcome. For example, an August 2020 report found that challenges to getting farmers and ranchers to adopt regenerative agriculture in the United States include insufficient awareness, knowledge, or willingness to change on the part of farmers and ranchers; the rise of farmland prices in recent years; minimal trusted technical assistance combined with unclear guidelines for how to implement regenerative practices; a lack of financial capital and incentives; the entrenched nature of the current agricultural supply chain; lack of awareness among key stakeholders; not enough truly independent research and science on regenerative agriculture; and an unfavorable political landscape. Granted, the report provides key levers and opportunities to overcome these barriers, but it's undeniably a tall order. (In my view, eating better meat would also mean eating less meat—though that's not necessarily a bad thing, given all the health problems associated with eating too much animal protein.)

Another challenge posed to the promise of better meat is that it might not be as environmentally friendly as its proponents make it out to be, even when produced using regenerative grazing practices. A report called "Grazed and Confused?" by the Food Climate Research Network at the

University of Oxford found, based on a review of the peer and non-peer reviewed literature, "The contribution of grazing ruminants [like cows, sheep, and goats] to soil carbon sequestration is small, time-limited, reversible and substantially outweighed by the greenhouse gas emissions they generate."[7] As we have seen, advocates of regenerative agriculture point to evidence suggesting otherwise. But the reality of the situation is that because the field is so nascent, there are few empirical studies available—this means that it's hard to make heads or tails of it all. Moreover, monogastrics like chickens and pigs cannot graze on grass (only the few bugs, grubs, and seeds within) and therefore their diet must be largely grain based. As food historian James McWilliams pointed out in an op-ed in the *New York Times*,[8] even if the current number of chickens and pigs were moved onto pasture, their feed would still most likely come from industrial farming, "thereby undermining the benefits of nutrient cycling."*

In direct response to McWilliams, Joel Salatin, one of the pioneers of regenerative agriculture, notes in an op-ed in *Grist*, "The land base required to [grow the] feed . . . is no different than that needed to sustain the same animals in a confinement setting,"[9] and therefore, in light of all the upsides of regenerative agriculture, we should advocate for it. As David Bronner (of Dr. Bronner's soap fame) argues on his company website—which is a surprisingly good source of information on regenerative agriculture—it remains to be seen whether this "monogastric loophole"[10] can be closed with an influx of inexpensive, high-quality feed grown in a regenerative fashion. Farmer and author Simon Fairlie counters that rather than close this loophole, we should focus on farming animals that do not need grains grown especially for them. He has said that cow and pig farming are the best choices, since cows subsist on grass that grows where other plants might not thrive, and pigs have an appetite for food waste, which means they can eat our leftovers.

Similar questions have been raised around the healthfulness of better meat. As journalist Tamar Haspel explores in the *Washington Post*, while

---

* As the Savory Institute notes, "As one organism consumes what it needs to live and grow, it takes nutrients from the environment and changes them in a way that is beneficial to its needs. When that organism (or its waste) becomes a food source for another organism, those nutrients are cycled to the next stage. The nutrients must keep cycling and recycling through the system."

grass-fed beef has less fat overall in comparison to industrial meat, and is higher in some vitamins, minerals, and omega-3s, their amounts are too small to be significant, or evidence of their value is ambiguous. Haspel said that for these reasons, one researcher she interviewed was "concerned that a reputation for healthfulness will make people believe that it's better for them than it is, which will lead to overconsumption."[11]

Skeptics of regenerative farming point out that rather than bringing grazing animals onto farms in order to advance our environmental goals, it may be more prudent to convert the vast expanse of existing pastureland into wilderness. In support of this idea, British environmentalist George Monbiot notes in the *Guardian*, "Not only would this help to reverse the catastrophic decline in habitats and the diversity and abundance of wildlife, but the returning forests, wetlands and savannahs are likely to absorb far more carbon than even the most sophisticated forms of grazing."[12]

Others who do see the value of regenerative agriculture but imagine alternative models propose that farmers either "partner" with animals in service of regenerative agriculture (let the animals eat the grass and spread their manure but also allow them to live out their lives) or rely on veganic agricultural practices (where no animal manures are used) and still succeed regeneratively. This latter model describes Woodleaf Farm in eastern Oregon, which I had the opportunity to visit in the summer of 2019. Veganic farmer Helen Atthowe, who cofounded the farm with her husband, Carl Rosato, told me that she is "really surprised people aren't using [veganic agricultural practices] because we know they are such a good way to build organic soil, organic matter, and to provide nutrients."[13] Skeptics of these alternative systems argue that they either are not profitable and therefore not scalable, or that they simply don't work ecologically—despite Helen's claims to the contrary. But thirteen years after Charles Martin, assistant professor at New Mexico State University's Sustainable Agriculture Science Center, explained that the benefits of veganic farming "haven't been demonstrated one way or the other, either economically or from a quality standpoint,"[14] this is still true—which doesn't bode well for its potential.

But at the end of the day, whether someone is for or against better meat may come down to a moral question: Is it permissible to take a life?[15] Those who renounce the killing of animals altogether will categorically balk at better meat, claiming animals raised explicitly for the purpose of being slaughtered are better off not being born at all. Yet proponents of better meat typically feel slaughtering animals is not only justified but a moral imperative, provided they had a positive life before their final moments.

Even if we do accept that raising animals for food involves killing them, animals can still suffer even on the happiest farms. Will Harris is unusual in the lengths he goes to in order to make sure the animals on his farm have very good lives. While White Oak Pastures avoids practices like separating calves from their mothers, dehorning young cattle, and castration, these practices remain common even on so-called humane farms. Free-range hens are often from breeds that grow to grotesque sizes at alarming rates, and male chicks born to laying hens are still routinely killed at birth because they offer no utility to the farm. "Humane" is not strictly defined, and thus allows for practices that seem quite cruel, even if we think animals raised this way have lives ultimately worth living. There's also a related worry that the mere existence of better meat can cause people to think that meat production generally is becoming more ethical, which of course isn't true. Another concern is that even if consumers purchase better meat at the supermarket, that does not mean they won't be eating plenty of conventionally produced meat at restaurants, while traveling, or at others' homes. Even some self-described "humane omnivores" confess to doing this. Furthermore, some animal advocates believe that humane treatment of animals who are raised for food is simply never going to happen: Legal scholars Anna Charlton and Gary Francione write, "Given that animals are property, and we generally protect animal interests only to the extent that it is cost-effective, it is a fantasy to think that 'humane' treatment is an attainable standard in any case."[16] I believe White Oak Pastures disproves this, but there have been enough cases of abuse at "humane farms" to see validity in their argument.

Here's the bottom line: While better meat is possible, it isn't easy. Making sure all of your meat came from well-treated animals requires discipline and vigilance. Whether it is zero animal meat or better meat that is ideal, proponents of the latter rightly point out the overwhelming majority of people still want to eat meat, and therefore, whether we like it or not, we might as well offer something better than industrial meat.

Except, as we'll see in the next chapter, does meat necessarily have to come from animals?

11

Plants 2.0

I'm eating a delicious burger. It's hearty, juicy, and has that distinctive umami flavor. Yet it isn't made from animal flesh; it's made from plants.

This is not your father's veggie burger. This isn't a soggy bean burger you find in the frozen section at the supermarket. I'm inside the headquarters of Beyond Meat, a Los Angeles–based producer of plant-based versions of everything from ground beef to pork to sausage.

Sitting in front of me is Ethan Brown, who founded Beyond Meat in 2009. Besides sporting a hat and a T-shirt with a cape-wearing cow on it, Ethan defies nearly every vegan stereotype: He's got the body of a linebacker, the height of an NBA point guard, and the kind of booming voice you typically hear narrating movie trailers. He explains that he grew up in Washington, D.C., but spent a lot of time with his father in the countryside on the family milk farm. As a kid he ate a lot of meat; his favorite sandwich was the "Double R Bar Burger" from Roy Rogers—a hamburger topped with a melted slice of American cheese and a piece of seared ham. Then somewhere in his teenage years, while pondering a career as a veterinarian, Ethan began to think about his relationship to the animals he ate. In his late twenties, while working on hydrogen fuel cell technology to help mitigate climate change, he became obsessed with answering a simple question: "Does meat need to come from an animal?" As he began to unpack that question, he got some pretty interesting answers, beginning with the composition of meat. He explains:

> Meat is essentially five things: it's amino acids, it's lipids, it's trace minerals, it's vitamins, and it's water. . . . All of those things are available outside the animal, they are not exclusive to the animal. What the animal is doing is using their digestive system and their muscular system to basically convert vegetation and water into muscle or meat. What we're doing is starting with that same source, we're starting with all plant-based material and we're starting with water, and we're using our system [which involves heating, cooling, and pressure] instead of the digestive system and the muscular system . . . to assemble those core materials in the architecture or against the blueprint of animal protein or meat, and so essentially building meat directly from plants.[1]

While I share Ethan's enthusiasm for plant-based alternatives to traditional animal food, it's a stretch to say he's *actually* turning plants into meat. Beyond Meat is making plant-based proteins *taste* more like meat protein, but its cutting-edge tech cannot, say, digest grass and actually transform it into animal body cells. If they were able to do that, someone who was allergic to pea protein but not to animal flesh would be able to eat Beyond Meat's products despite pea proteins being used as inputs. This is not the case. Someone who is allergic to the ingredients Beyond Meat uses will likewise be allergic to their products, no matter how much it looks and tastes and smells like meat. Of course, there's nothing wrong with that.

When I ask Ethan to explain the rationale behind plant-based meat, he explains:

> I couldn't find anything else that I could focus on where you could simultaneously advance four solutions by focusing on one thing. . . . These solutions are in regard to human health . . . heart disease, diabetes, and cancer, particularly in African American and Latino populations there are higher levels, and then you look at climate, and it's controversial around climate about how much livestock really contributes, but I think we can all agree it's a really significant contributor. I think of the natural resources, the amount

of water, and land, etc. And then animal welfare—the obvious one. Focusing on all four of those things, moving all four of those forward at the same time is something I think is unique to this solution and company.

Advocates of plant-based eating have rhapsodized for years that human health, animal welfare, and the environment will all improve dramatically if everyone were to adopt a plant-based diet, so Beyond Meat is certainly not alone in touting these claims. But as Beyond's popularity grows—settling in at about a $9 billion valuation as of July 2021—its influence has grown correspondingly outsized as it has signed partnerships with KFC and McDonald's. In terms of the environmental benefits, Ethan cites a Beyond Meat–commissioned study by the University of Michigan, which found that the company's products require 99 percent less water, 93 percent less land, and about half the amount of energy—plus their production involves 90 percent less emission of greenhouse gases compared with an industrial burger. "What fires me up so much about this is that you can make so much change by just changing the protein that is in the center of the plate," he explains. "You don't have to do anything, really, other than just say okay I'm either going to have [a meat] burger or I'm going to have a Beyond Burger, [and then] pick the Beyond Burger, [and] it tastes great, you feel great, it's better for your health, and it advances all of these goals. Pretty exciting."

Jazzed by my conversation with Ethan and the promise of plant-based meat, I go on a tour to learn more about how Beyond Meat's products are made. What I discover quickly reminds me of the work of taste optimizers like Howard Moskowitz, trail blazers who used fancy gadgets to guide their search for the ideal levels of sweetness, saltiness, and other features in food products. It seems everyone is eager to make their foods as tasty and familiar as possible, and those who want to replace meat with plants are fully aware they cannot count purely on the public's moral sentiments to succeed; they must convince our taste buds as well.

First, I meet Parker Lee, a scientist in the "analytical lab." His team's mission is, among other things, "to understand the texture, fat binding, and general properties of the products . . . [they] make to get them close to animal meat, if not exactly the same." He first shows me a machine he calls "the e-mouth," a texture analyzer used to electronically simulate the human mouth and compare the force required to chew animal-based meat and plant-based meat. "This allows us to see how close we are, what improvements we're making with different textures or different additions, different structuring of the material via how we formulate it, and get an actual scientific read-out, not someone just saying this one is chewier than that one,"[2] he continued.

Lee then places a Beyond Burger in the machine and presses some buttons that produce video-game-like sounds. A moment later a steel rod presses against the burger, and a display records the amount of force exerted. Fat drips from the burger and along the side of the machine; Lee tells me he has "other testing parts we can put it under and see how much fat is actually released and measure that quantitatively . . . [which] is important in developing our textures, our formulations, even cooking times." He adds, "We optimize everything to try to get the consumer the feeling of eating animal-based meat."

Next, he shows me the "meat microscope," a multiuse, high-powered electron microscope that allows his team to visualize the structure of plant- and animal-based meat. Pointing at images on the screen corresponding to each kind of meat, he tells me, "The protein fibers are in white, and the fat pockets are in these darker colors. . . . It shows us that we are pretty close [to replicating the structure of animal-based meat] . . . but we are always working on improving it."

Next, I meet Daniel Ryan, who is the director of chemistry. Among other tasks, Ryan works to understand the flavor of animal-based meat and uses that knowledge to mimic the flavor in a plant-based product. Using what he calls an "e-nose machine," a gas chromatography–mass spectrometry instrument (GCMS), he tries "to look at those flavor and aroma components [in animal-based meat] and understand which are the important ones and how can we find them in the plant kingdom and use them in a plant-based meat."[3]

To illustrate how it's done, Ryan takes a pipette and soaks up juice from a cooked beef steak and squirts it into a vial. He then places the vial into the GCMS, which heats the liquid to a cooking temperature of 165 degrees Fahrenheit, where the aroma compounds can be analyzed. Then Ryan presents me with two vials that contain liquids derived from plants. The smells are hard to describe—one has a fatty, oily smell, while another has hints of the smell of fur—but collectively they smell like a traditional hamburger.

The last stop on the tour is Julie Wushensky, a research associate who primarily works on the color and appearance of Beyond Meat products. She is standing beside a cutting board filled with beets, and is surrounded by beakers of strawberries, blueberries, and raspberries. She and her team are "trying to replicate all of the visual cues that animal meats create for us in plant-based meats."[4] She says, "We are very consumer oriented, so we are trying to keep a very clean label, so we source all of our colors from natural ingredients like fruits, vegetables, edible flowers, that sort of thing." I ask her to explain how they extract the color. "We primarily use beets for their very meaty, almost blood-like color. . . . You can chop them up, you blend them a little bit, you filter press them to get rid of the starches and the fibers and the skin of the beets and then you can concentrate it, which you can either do by getting rid of the water content or doing very basic processes like that."

Beyond Meat is one of many innovative companies that produce meat from plants, rather than the traditional way of cycling plants through animals to make meat. Will redefining meat in this manner finally put factory farming and the myriad problems associated with it in the dustbin of history?

One hurdle is an immediate and literal one: Many cattlemen and related industry groups want to prevent companies like Beyond Meat from referring to their products as "meat." While the name "Beyond Meat" clearly suggests a product other than meat, many cattlemen's associations and their allies still see this as too close for comfort, so they are lobbying Congress to exclude plant-based products from the "meat" label. Similar efforts have been made by the dairy industry to forbid makers of soy, almond, oat, and other plant-based milks from calling their products "milk." The European Union Parliament put forth a proposal in April 2019 to require that veggie burgers be marketed under an alternative name, such as "veggie discs," with veggie sausages labeled something like "vegetable protein tubes." In October 2020, that proposal was voted down, but a proposal to expand a ban on terms such as "yogurt-style" and "cream imitation" for non-dairy replacements did pass, extending previous limitations on the use of words like "milk" and "butter" on vegan alternatives. Similar laws regulating the names of veggie products have been passed in the United States, such as in Missouri and Arkansas. As of writing, they are currently being contested by advocate groups including the Animal Legal Defense Fund and American Civil Liberties Union (ACLU). (Interestingly, the major meat processors aren't taking sides, likely because they themselves are invested in plant-based products. Industrial animal ag companies don't seem to care what they sell as long as they are making money, whereas cattlemen have been raising animals for food for decades, and for economic and cultural reasons, are more hesitant to accept this kind of change.)

Whatever plant-based meat is ultimately called, cost remains the major hurdle: Products like the Beyond Burger are more expensive than animal-based meat. At many restaurants where it's served, including Burger King, the Impossible Burger, made by Impossible Foods, is about $3 more than an animal-based burger. Carl's Jr. sells a Beyond Burger that similarly costs as much as $3 more at some of their locations. But as these companies continue to streamline and scale production, prices will invariably come down. (In fact, in January 2021, Impossible Foods cut prices on average about 15 percent for food-service distributors that sell to restaurants; the following month, the company slashed its suggested prices for US grocery stores by 20 percent. Not long after, Beyond Meat began offering its most aggressive promotional retail pricing so far.)

But even if plant-based meat can reach price parity with animal-based meat, another question remains: Is plant-based meat any healthier? Plant-based meat is often loaded with salt and saturated fat. Consider the Hot Italian Beyond Sausage. It's made from a laundry list of ingredients including pea protein isolate, refined coconut oil, sunflower oil, natural flavor, rice protein, fava bean protein, potato starch, salt, fruit juice, vegetable juice, apple fiber, methylcellulose, citrus extract, and calcium alginate casing. As for its nutritional information, one cooked link has 500 milligrams of sodium and 5 grams of saturated fat, or 21 percent and 25 percent of the recommended daily value, respectively. Former US Agriculture Secretary Dan Glickman said of Beyond Meat, "We can't really market it . . . as necessarily better for you, because we don't know."[5] The Impossible Burger has less sodium than the Beyond Sausage, at 370 milligrams per patty, but it has 8 grams of saturated fat. Food reporter

Lauren Wicks writes, "While seeing the Impossible Burger on a menu can feel like a godsend to vegans and vegetarians at a restaurant, this burger should be treated as an indulgence in the same way a healthy omnivorous eater would view a standard cheeseburger. It's fine on occasion, but the Impossible Burger shouldn't be seen as a go-to meal for health purposes."[6]

Perhaps in response to comments like this, in May 2021 Beyond Meat debuted a leaner patty with 35 percent less saturated fat. (Impossible Foods had made a similar update to its burger in December of 2019, decreasing its saturated fat by 43 percent.) But this too does have some downsides. The reason Impossible and Beyond products have taken off in a way that earlier, vegetarian-targeted brands such as Boca and Gardein never quite did is not their nutritional profile. It's the fact that they actually taste good. The great opportunity of these innovative, super-realistic meat alternatives is their potential to convince die-hard meat eaters that they can, in fact, eat satisfying and delicious food without relying on animals. Making comfort food healthier at the expense of taste decreases the chances that consumers will make the switch. Then again, there is an argument to be made that while most consumers don't *actually* want to eat a healthy diet, those who are purchasing Impossible and Beyond products do, and so then these companies would be striking an appropriate balance with respect to their target audience.

One question that invariably arises: To *what*, exactly, are we comparing the nutrition of these products? Sure, everyone would be healthier if they ate solely whole plant-based foods, but it's unrealistic to expect the masses to swap out meat, whether plant-based or otherwise, for fruits and vegetables. (Would you be surprised to learn that the salad on Wendy's menu isn't their most popular item? I'm glad they offer it, but nobody goes to McDonald's for the fruit. They go there for greasy, good old-fashioned Americana. As we have seen time and time again, the things we humans like—lots of fat, salt, sugar, etc.—aren't good for us.) If people are going to eat meat, the argument goes, it is better nutritionally (not to mention environmentally and ethically) when it comes from plants instead of animals. Beyond Meat notes that in comparison to a leading brand of brat-style pork sausage, their sausage has 38 percent less saturated fat and no "hormones, nitrites, nitrates, soy, or gluten . . . either." And indeed, in support of these claims, a 2020 study conducted by researchers at Stanford

Medicine found that TMAO levels—a molecule linked to cardiovascular disease risk—were lower when study participants were eating plant-based meat as compared to study participants who were eating animal-based meat. So, while perhaps not as healthy as kale, the available evidence suggests plant-based meat may be healthier in some respects than animal-based meat.

Some skeptics of plant-based meat also claim it isn't as environmentally friendly as it seems. Indeed, some of the ingredients in plant-based meats do come from monocultures (fields that grow solely one crop, often to the detriment of soil biodiversity). While monoculturing is very efficient, the lack of diversity in crops can lead to various problems, including the spread of plant-destroying diseases and soil nutrient depletion.

As food writer Alicia Kennedy explains in *In These Times:*

> Here's the flaw. Transitioning away from beef will not be enough to deal with the broader ecological crises we face, from climate change to biodiversity loss to resource scarcity. The Impossible Burger is mostly soy and the Beyond Burger is mostly pea protein. By producing these burgers on a massive scale for cheap, these companies are supporting a plant-based food system—but one that relies on monoculture rather than a diversified crop system. Diversifying crops is crucial to soil and wildlife health and helps keep small farmers in business. . . . Crop diversification sequesters more carbon, cycles nutrients more efficiently, reduces nutrient runoff and allows soil to retain more water—all of which makes farms more resilient to extreme weather. By relying primarily on soy and pea, Impossible and Beyond are doubling down on a harmful system.[7]

In response, plant-based meat advocates contend that the positives from eliminating our dependence on industrial meat would vastly outweigh any negatives brought by monoculturing. Additionally, factory farms already depend on monocropping to keep costs down—a practice that could be more easily ended once the majority of plants are grown to feed humans instead of farmed animals. In other words, we can make improvements to the environmental footprint of plant-based meat without using animals.

What about fears over the reported use of genetically modified organisms (GMOs) in plant-based meat? In 2019, Pat Brown (no relation to Ethan), the founder and CEO of Impossible Foods, tackled that controversy directly by posting an article in defense of their use of geneti-cally modified soy protein. Insisting the science has not shown any prob-lems with GMOs,* Pat argued their plant-based burgers are healthier and more environmentally friendly than the animal products they replace. This did not go over well with the regenerative agricultural community, and some proponents of plant-based meat—including those who support the use of GMOs—have argued that Impossible Foods could have done more to avoid linking the brand to the controversy. Impossible Foods is large enough that its ingredient needs can shift farming practices—there may not be enough non-GMO American soy to satisfy America's hunger for Impossible Burgers right now, but Pat could have announced they were accepting GMO soy temporarily as they transitioned toward only using non-GMO soy—essentially forcing farmers to switch their practices or lose lucrative contracts with Impossible Foods.

What about people like my parents, who will just always prefer "the real thing" no matter how well Impossible Foods, Beyond Meat, and other companies market their products? Indeed, a 2019 study found that among 987 participants, 25.3 percent were "not at all likely" to eat plant-based meat. These numbers may be changing: A study by the Plant Based Foods Association and the Good Food Institute commissioned with SPINS found US retail sales of plant-based meat grew 45 percent in 2020 (twice as fast as conventional meat), hitting $1.4 billion, up from $962 million in 2019. But even with that growth, plant-based meat still only accounts for about 1.5 percent of total retail meat sales in the United States. (This figure is for sales, not volume, and since plant-based meat tends to cost more than animal-based meat, the actual volume of plant-based meat that Americans are eating is likely a good amount lower.) As journalist Larissa Zimberoff points out in her book *Technically Food*, "Because of the near-constant headlines, we think that plant-based is killing it when in reality it has captured just [a tiny] percent of the market. While plant-based meat saw phenomenal traction in supermarkets due to changes in our shopping

---

* The use of GMOs is widespread. As the FDA explains on its website, "It is very likely you are eating foods and food products that are made with ingredients that come from GMO crops. Many GMO crops are used to make ingredients that Americans eat such as cornstarch, corn syrup, corn oil, soybean oil, canola oil, or granulated sugar. . . . Although GMOs are in a lot of the foods we eat, most of the GMO crops grown in the United States are used for animal food."

habits during the [COVID-19] pandemic, it will take immense efforts to compete with meat."[8]

As low as that 1.5 percent of market statistic may seem, it likely still has an impact on the number of animals raised for food, thereby cutting into demand for conventional meat. And more importantly, it proves that it's possible for people to enjoy—and even prefer—plant-based meat. Given the environmental and ethical advantages of plant-based meat over industrially raised animal flesh, it's certainly plausible to imagine that 1.5 percent increasing, especially if more people accept that something needs to be done about climate change, chronic disease, and unnecessary animal suffering.

Beyond Meat recently went public. While its stock was originally priced at $25, it nearly quadrupled within a month—despite the company not yet generating a profit—and peaked at $239 in July 2019, settling at about $100 as of November 2021. Impossible Foods hasn't yet gone public as of this writing, but it has been raising a lot of money, including $300 million in a fund-raising round in May 2019, with backing from such celebrities as Serena Williams and Katy Perry, and another $200 million in August 2020. This shows there is lot of optimism about alternatives to animal flesh ultimately succeeding. It's not unreasonable to compare Ethan Brown and Pat Brown to major figures in the history of American food production like Gustavus F. Swift, Jesse Dixon Jewell, and Phillip Armour. While the two are working against the legacies of these food-production and transportation innovators who brilliantly but disastrously intensified animal agriculture, Ethan and Pat are nevertheless similar in their ambition, excitement, and revolutionary fervor.

At the end of the day, as we've seen in this book, preferences, cravings, willpower, and even nutritional needs differ among people, and it isn't easy to imagine everyone willingly giving up animal meat for veggie proteins. That 1.5 percent will almost certainly expand, but if my experiences are any indication, there is a hard ceiling.

This raises the question: If "real meat" is what the majority of consumers want, might there be way for them to have their cows and eat them too?

12

Meat without Slaughter

In the palm of my hand is a $100 chicken nugget. No, it's not dusted with gold, infused with truffle oil, or crusted with Almas beluga caviar. Instead, it's made from a perfectly intact chicken who is probably still alive today. It's early March 2019. I'm at Eat Just, a San Francisco–based food manufacturing company. Founded in 2011 by Josh Tetrick and Josh Balk, this billion-dollar-plus company is known for creating an array of plant-based products, everything from mayonnaise made from peas to eggs made from mung beans. But for the past few years Eat Just has been working stealthily on a new product: cell-cultured meat. Also known as in-vitro, lab-grown, synthetic, cultured, cell-based, cultivated, and clean meat (or as sociologist Neil Stephens perhaps best described it, an "as-yet undefined ontological object"[1])—cell-cultured meat is real meat grown from animal cells rather than slaughtered animals.

According to Vítor Espírito Santo, associate director of Cellular Agriculture at Eat Just who gave me a tour of one of the company's labs earlier in the day, the process is complex in procedure, but fairly simple to understand in principle: Stem cells from an animal are grown in a nutrient-rich blend of sugars, amino acids, fats, proteins, oxygen, and water. After the cells multiply, they merge together, causing them to grow into small strands of muscle tissue. These strands are then layered to form familiar meat products that can be prepared and cooked like conventional meat—from turkey sausages to pork hot dogs.

Vítor explained how cell-cultured meat is produced in significantly more sanitary and controlled conditions than industrial meat (which is often loaded with antibiotics and synthetic hormones and swimming with bacteria like salmonella, campylobacter, and E. coli). He also highlighted how cell-cultured meat's environmental impact is potentially much lighter: For instance, a 2011 study estimated it could reduce land use by 99 percent, cut water use by 90 percent, and reduce greenhouse gas emissions by 96 percent. A joint study between the University of Oxford and the University of Amsterdam in the same year estimated cell-cultured meat would require 7 to 45 percent less energy to produce than pork, lamb, or beef, while requiring less land and water than chicken rearing. Finally, Vítor explained that the taste and nutrient profiles can be fine-tuned (such as swapping out bad fats for good fats)—and best of all, no blood is spilled.

It may seem like a novel idea, but the concept of cell-cultured meat has actually been around for decades. Back in 1930, the British secretary for India, Frederick Edward Smith, Earl of Birkenhead, wrote in *The World in 2030 A.D.*: "It will no longer be necessary to go to the extravagant length of rearing a bullock in order to eat its steak. From one 'parent' steak of choice tenderness, it will be possible to grow as large and as juicy a steak as can be desired. So long as the 'parent' is supplied with the correct chemical nourishment, it will continue to grow indefinitely, and, perhaps, eternally."[2]

One year later Winston Churchill predicted in an essay for *Strand Magazine* that within fifty years we would be growing edible animal parts to "escape the absurdity of growing a whole chicken in order to eat the breast or wing, by growing these parts separately under a suitable medium."[3] Churchill was referring to the parts of the chicken that often go to waste (feathers, bones, and gizzards) when we only eat breasts, drumsticks, and thighs. "The new foods will be practically indistinguishable from the natural products from the outset," he continued.

I was about to find out if Smith and Churchill were right.

According to VP of Product Development Chris Jones, who prepared my futuristic dish for me, more people have been to the moon than have eaten a cell-cultured chicken nugget. With the gravity of the moment in mind, I carefully turn the nugget upside down and right-side up again to inspect it further—it certainly looks like chicken. Then I take a whiff—it certainly smells like chicken. Finally, I take a small bite. After a moment of chewing, I say exactly what the chef had hoped for: "It tastes just like . . . chicken."

My mind is blown. The possibilities in this burgeoning field of "cellular agriculture" are seemingly endless. But to what extent is cell-cultured meat a viable option for ending factory farming and the problems associated with it?

Skeptics of cell-cultured meat (who Jim Mellon, British billionaire and author of *Moo's Law* comically refers to as "the agro Luddites"[4]) point to the obvious: It costs a fortune to create, making it impossible, for now, to scale up production and price it competitively with industrial meat.

Pat Brown of Impossible Foods is one of these skeptics. When an interviewer asked him why he doesn't get into cell-cultured meat instead of plant-based meat, he said, "The simple answer is because that is one of the stupidest ideas ever expressed. First of all, it's not true you can do a better job that way. Because then you buy into, at best, the same limitations that a cow has. And it's economically completely un-scalable. If we could grow tissues that were a meaningful replica of animal tissues at an affordable price from stem cells, it would revolutionize medicine. Look around you. It's not happening."[5] Of course, if cell-cultured meat ever were successful, it would be a huge competitor for Impossible Foods, so Brown may be somewhat biased here. That said, he has a point.

Proponents of cell-cultured meat concede that large-scale manufacturing of it—which includes the cost of cell lines, cell culture media, enormous bioreactors used to grow the meat, facilities, and skilled labor—won't come cheap. But they argue that advancements in recent years should inspire optimism. Consider that when Mark Post, chief science officer at Mosa Meat, created the first cell-cultured hamburger in 2013, it had a whopping price tag of $325,000; in 2015, it cost $44,000, and now, he claims it costs just a little more than $11 per burger (or $37 per pound). Similarly, a plate of UPSIDE Foods' (formerly Memphis Meat) cell-cultured chicken that used to cost tens of thousands of dollars to produce can now be grown for less than $50. In other words, while cell-cultured meat may still be pricey in comparison to a Big Mac, costs may continue to decline in the near future when (and if) production methods are streamlined. Then again, while cell-cultured meat companies have made numerous ground meat products—burgers, nuggets, patties, etc.—more complex cuts like steak and pork loin are only in the experimental stage (though it's not like plant-based steak is close on the horizon either). And given that whole cuts account for about 40 percent of beef consumption and most of the chicken that people eat, that's a major problem.

Even if cell-cultured meat is economically feasible, other skeptics doubt whether consumers will really eat it. Perhaps the thought of eating meat prototyped in a lab by a radical new method will seem weird and disgusting to them—the prejudice that so-called frankenfoods are unnatural. A 2019 study found, for example, that less than a third of participants in the United States indicated they were very or extremely likely to purchase cell-cultured meat. In response, many supporters of cell-cultured meat note that all processed foods—including fortified cereals, peanut butter, beer, and tomato sauce—start in a lab before they are mass-produced in a factory. Yet you don't hear consumers complaining about this fact with those products. Similarly, once cell-cultured meat is on the market, people won't care about its origins. In other words, if it tastes the same as conventional meat, it may not be long before more people start to think meat from industrial slaughter is more off-putting than growing meat in bioreactors.

A common critique of cell-cultured meat is that, like plant-based meat, it distracts from the conversation around the possibility of better meat from regenerative agriculture. But the environmental benefits of small-scale farming operations aren't clear cut, and the factory farm system exists precisely because our appetite for meat is so great. It simply would be impossible to feed billions of people around the globe with humanely raised animals. Proponents of cell-cultured meat argue the widespread adoption of synthetic meat may lessen (and eventually end) demand for industrial meat—which may in turn increase demand for better conventional meat among people who hold out for it.

Even if cell-cultured meat can thrive alongside regenerative agricultural practices, some environmentalists argue it's too early to deem it "eco-friendly." For example, Richard Young, policy director at the Sustainable Farming Trust, has referred to cell-cultured meat as "fool's gold,"[6] citing that the technology has yet to be fully evaluated in terms of the amount of energy it would require. Young isn't wrong. For example, a 2019 analysis of cell-cultured meat out of the University of Oxford argues that while it will help mitigate climate change in the near-term, it may actually accelerate it faster than industrial meat in the long run. But the authors of the study made two questionable assumptions in their models: (1) cell-cultured meat will continue to be produced using the same methods of energy generation currently powering production (fossil fuels), and (2) this will continue for an additional one thousand years. Those who take issue with these assumptions argue that it seems very unlikely we'll still be using the same energy-inefficient methods to make cell-cultured meat in the future. As journalist and author of *Billion Dollar Burger* Chase Purdy notes in *Quartz*, because no cell-cultured meat company has a fully operational production facility yet, "The reality is that nobody knows how much energy it takes to create cell-cultured meat at scale."[7] In short, it's difficult to see how cultured meat could be any more environmentally destructive than industrial meat, which is currently estimated to account for 20–33 percent of global freshwater consumption, 80 percent of agricultural land use, and 14.5–18 percent of the world's greenhouse gas emissions.

In a similar vein, some environmentalists and health advocates are concerned cell-cultured meat will unleash "synthetic biological creations"[8] that will harm people and the planet. As Dana Perls, senior food and technology campaigner at Friends of the Earth, expressed in a report, "The makers of this new wave of products claim to minimize the environmental impact of industrial factory farming by replacing animal products with synthetic products; but these engineered substances may also be resource intensive in their use of energy and water, as well as using feedstocks, like sugar and methane, and chemicals. The new animal replacement products are being marketed or promoted as 'clean meat,' 'animal-free,' 'plant-based' or 'climate-friendly,' but with questionable substantiation."[9]

Perls further points to examples in history where products were recklessly rushed to market in order to solve some urgent problem without thinking about the long-term consequences, such as with the now-infamous introduction of the insecticide DDT. Although many proponents of cell-cultured meat agree it would be helpful to establish additional safety protocols and ensure they are enforced, they take issue with some of this language for being scientifically inaccurate (cell-cultured meat doesn't necessarily rely on genetic engineering or synthetic biology, for example) and too hyperbolic. Even though cell-cultured meat has not been created at scale, we have enough theoretical and observational knowledge to confidently say it will not continue replicating indefinitely and overrun the world like the Blob. Any new technology poses risks, of course, and the potential harms—however alarming they may be—should be weighed against the potential benefits of using the technology. This reminds me of what Ruth DeFries, author of *The Big Ratchet: How Humanity Thrives in the Face of Natural Crisis*, told me: "Every technological solution that humans have come up with is ingenious but then leads to some other kind of problem. . . . Our work . . . is to think about our technologies and to think about the solutions we come up with [in such a way that we] can minimize those backlashes."[10] Indeed, despite disasters like Chernobyl and Fukushima, for example, many still support the development of nuclear energy as a way to reduce our reliance on fossil fuels.

Proponents of cell-cultured meat also note that the founders of these companies are extremely mission driven and are therefore unlikely to use their technology irresponsibly. But this rosy sentiment doesn't quell the fears of many cell-cultured meat critics who argue that technology doesn't live in a vacuum. Pointing again to the consequences when massive profit-driven companies get their hands on transformative food technologies—even those initially developed by well-intentioned folks (think of the many controversies surrounding Monsanto and GMOs)—critics note that industrialized animal agriculture companies may someday leverage cell-cultured meat technologies to fatten their wallets unchecked, which could come at the expense of our health and the environment. Indeed, this may already be happening, as some industrial meat producers, such as Tyson and Cargill, are now investors in cell-cultured meat companies. But for many advocates of cell-cultured meat, corporate involvement is exactly what they want. As reporter Elaine Watson explains in *Food Navigator*, "Meat companies are obvious partners in that while they are no experts in cellular agriculture, they already have an established infrastructure for handling, packaging, distributing, selling, and marketing meat."[11]

Another concern coming from consumer advocacy groups is not with the technology per se, but rather the lack of transparency involved in using it. They argue that because the processes these companies are using are proprietary, consumers are essentially being asked to blindly trust them. While it's true that much of the intellectual property is protected by patents, cell-cultured meat products will actually have to undergo more scrutiny before hitting the market, as they will be regulated by *both* the USDA and FDA. Additionally, as Mike Selden, CEO of cell-cultured fish company Finless Foods explained to me when I visited his lab, it's incredibly difficult to argue that cell-cultured meat could possibly be any less transparent than the alternative, given that factory farms are so protected there are literally laws preventing people from seeing and sharing what goes on inside. The mere fact that I was welcomed by all of these companies and given tours of their facilities is quite significant when you think about what would have happened to me if I tried to do the same at a factory farm.

There is also legitimate worry that the triumph of cell-cultured meat could lead to greater consolidation in the food industry. Growing cell-cultured meat will remain an extremely pricey endeavor for the foreseeable future. Even if some of the companies working on it now are comparatively small, they may well be gobbled up by conglomerates in the future. (Indeed, there is already a precedent for this in the plant-based meat space: In April 2021, for example, JBS, the world's largest processor of fresh beef and pork, entered in an agreement to purchase Vivera, Europe's third-largest plant-based food company.) "Mimicking the meat giants, aspiring to their mammoth production and monopolistic tyranny, is to build a 'new' industry in the image of the world's ugliest business,"[12] writes journalist Charlie Mitchell in the *Baffler* in opposition to this possibility. Jenny Kleeman, author of *Sex Robots and Vegan Meat*, also has concerns about the influence of giant meat producers in cell-cultured meat, writing in the *Guardian* that if this happens "we will become ever more dependent on remote corporations with highly specialised technology to meet our basic needs."[13] Another concern is that even if they remain independent, these companies may become the next Tyson or Nestlé in their own right. Indeed, Errol Schweizer, former vice president of grocery for Whole Foods laments to host Nil Zacharias on an episode of the *Eat for the Planet* podcast about the corporatized direction alternative protein companies are headed: "This does not get us to what we need, which is food sovereignty, which is the control of the food system by the people. . . . This isn't what we set out to do. We've become who we despise."[14] (This of course assumes cell-cultured meat companies succeed.)*

In the meantime, another sort of mistrust has been growing, leading to skepticism that these companies will ever be able to mass-produce their cell-cultured dreams at reasonable prices. Part of that skepticism has nothing to do with cell-cultured meat companies at all—there is a long tradition of Silicon Valley startups generating hype and massive investments with little to show for it. Theranos is a notorious example. For years, the company, led by its mercurial founder, Elizabeth Holmes, claimed to be improving blood testing when in fact the only thing it was actually good at was *pretending* to improve blood testing.

But cell-cultured meat companies can also blame themselves for some of the skepticism directed their way. As journalist Tom Philpott quips in *Mother Jones*, "So far, the industry has proven far more adept at churning out optimistic timelines than it has at offering consumers an alternative to meat from animals."[15] For example, Eat Just (then known as Hampton Creek), predicted in 2017 that they would have cell-cultured meat in stores by the end of 2018—and they continued to predict this for much of 2019 and 2020. (CEO Josh Tetrick was far from the only cell-cultured meat developer to make these sorts of overly optimistic claims—Finless Foods CEO Mike Selden told *Food Navigator* "by the end of 2019 we plan on having price parity with Bluefin tuna."[16]) It wasn't until December of that year that they sold their first product under the brand GOOD Meat at restaurant 1880 in Singapore, after the country's food agency approved the sale of cell-cultured meat. (While the FDA and USDA released newly agreed-upon standards for the regulatory oversight of cell-cultured meat in 2019, they still haven't granted approval to any company to sell their products in the United States.)

As part of the launch, on the menu for about $17 was a trio of sample dishes: bao bun with crispy sesame cultured chicken and spring onion;

---

* To those who balk at Big Ag as a whole, Breakthrough Institute's Ted Nordhaus and Dan Blaustein-Rejto put it like this in *Foreign Policy*: "Any effort to address social and environmental problems associated with food production in the United States will need to first accommodate itself to the reality that, in a modern and affluent economy, the food system could not be anything other than large-scale, intensive, technological, and industrialized. . . . A better food system will build on these blessings, not abandon them."

phyllo puff pastry with cultured chicken and black bean puree; and a crispy maple waffle with cultured chicken with spices and hot sauce. GOOD Meat remains on the 1880 menu today. (According to research done by Eat Just in cooperation with a leading management consulting firm, 70 percent of Singaporeans who have sampled GOOD Meat said that it tasted as good or better than conventional chicken, and nearly 90 percent of those diners said they would substitute conventional chicken with cultured chicken.) While this is a historic milestone, Tetrick told CNBC, "We didn't set it at a cost where we're making money,"[17] which suggests they are still far from widespread commercial viability. And although they did make the product available in a few other restaurants in Singapore in 2021, it's not clear if their plan to be in retail stores by mid- or late 2022 will be another deadline Eat Just cannot make. Only time will tell.

Ben Wurgaft, author of *Meat Planet: Artificial Flesh and the Future of Food*, claims that the reason why cell-cultured meat is perpetually "five years away"[18] from restaurants and grocery stores is that this framing is useful for securing funding from venture capitalists, who don't want to invest in projects with longer time horizons. Fudging the timelines to get more funding may seem like a necessary strategy for now, but it's a risky gambit; after enough blown deadlines, these companies could come out looking deceptive or even fraudulent. At this stage, with the mass producibility of these products yet to be proven, misbehavior from even one cell-cultured meat company could potentially damage the reputation of them all. As author Martin Rowe warns:

> It is possible, perhaps even probable, that cellular agriculture is hosting a Theranos "unicorn" among its start-ups: where the promise of a technology that changes everything (and doesn't exist) is delivered by a young, charismatic founder who has considerable corporate investment and star backers, and whether deliberately or not misleads everyone to the detriment of those trying to do similar work in that space. A culture that encourages anyone to start their own business; that fails to police the hype; and that dampens due diligence, corporate responsibility, and realistic timeframes and actual deliverables to reward charismatic leaders or media-friendly funding pitches is one that opens itself up to potential fraud and recklessness.[19]

While this may come across as alarmist, it might also turn out to be too optimistic. Perhaps there will be even more than one cell-cultured meat company that deserves this comparison to Theranos.

However, fund-raising is not the only explanation for the overhype. In a talk at a 2018 conference hosted by New Harvest, a nonprofit working to advance cellular agriculture, technologist Adam Flynn offered a perhaps more charitable theory to explain the so-far unfounded optimism of cell-cultured meat companies. The problem, Flynn said, is that these companies are run by biologists instead of engineers. The former believe their job is figuring out how to create meat from the cells of animals; but the latter would recognize the real task at hand: not to create cell-cultured meat, but to create *machines* that create cell-cultured meat. (Christie Lagally, a former Boeing engineer and founder of Rebellyous Foods, a food production technology company working to make plant-based chicken price competitive with traditional chicken products, has made a similar argument about the need for "equipment that is made for doing what we're doing rather than using old equipment that was made for something else.")[20]

On the flip side with respect to price parity, there's the concern over the fact that technological progress increases the efficiency with which a resource is used, thereby fueling demand and boosting more consumption of that resource. This scenario, when the rebound effect is greater than 100 percent, exceeding the original efficiency gains, is called Jeavons paradox. A classic example is lighting, which has become vastly cheaper as the world has moved from lamp oil to tallow candles to incandescent bulbs to fluorescent and LED bulbs. As a result, people use more lighting than ever before. This same logic can be applied to cell-cultured meat. If cell-cultured meat is produced very efficiently it could become very inexpensive, and if it becomes very inexpensive, people will be able to consume even more of it. If that happens, many of the benefits we hoped to see from cell-cultured meat—such as improvements in human health and greenhouse gas emissions—could be undermined.

Proponents of cell-cultured meat in turn point to the many experts who think the actual rebound effect would be very small and that other measures would limit consumption. There is a limit to how much people can eat, after all, and a limit to how much meat they desire. Even if cell-cultured meat became extremely affordable, it's unlikely most people would want to shift to a purely carnivorous diet, and it may even be unlikely most people would want to eat significantly more meat than they currently do (with the exception of low-income countries).

Meanwhile, could the introduction of cell-cultured meat spell doom for small, local farmers who are already struggling to compete with industrial farms? As discussed in *AgFunder News*, while some in the field note they are "not out to transform family farms" and "there is a place for sustainable high welfare operations,"[21] others take a darker view, not having "any illusion[s] that smallholder farms are safe." As one farmer and meat-processing specialist pointed out, if cell-cultured meat does become cheaper than better meat, "it takes the choice away from the consumer. If you can make it cheaper, then you've changed the whole system." This may be true, but there will always be at least some consumers fixated on naturalness who will be willing to pay more for what they believe to be a superior product.

But even if a massively scaled-up cell-cultured meat industry eats into some demand for higher welfare meat and small family farmers' jobs are put at risk, there is still a dimension along which the expansion of this industry and the resulting consequences for factory-farming jobs would be quite good. If cell-cultured meat scales dramatically, the vast majority of the jobs being replaced will be those primarily held by vulnerable immigrants and other workers whose safety and rights are routinely violated by the industrial meat producers who employ them. As Molly Anderson, professor of food studies at Middlebury College told me, sexual and ethnic harassment, verbal and emotional abuse, low wages, scant benefits, high rates of illness and injury, abysmal housing and working conditions, and inhumane line speeds have all been documented by groups like Oxfam America and Human Rights Watch. "[Some] poultry workers in processing plants are not even allowed bathroom breaks, so they have to wear diapers for work,"[22] she explained.

Max Elder, former research director for the Institute for the Future and a cofounder of plant-based meat company Nowadays, noted in the *Routledge Handbook of Animal Ethics* that it seems likely cell-cultured meat manufacturing jobs could only be better by comparison. Commenting on the abysmal working conditions in industrial slaughterhouses, Elder said, "While large-scale cultured meat production has not yet been realized, it is certainly possible that this higher-skilled labor and cleaner production process will enable a new industry with higher labor standards and better wages. Some have imagined these facilities as located in urban centers, open to the public, and much more transparent than current animal production. Imagine the fun you've had at your local brewery, but replace the beer with meat, and you get the idea of one vision for what this might look like. It's by no means inevitable, but it is possible."[23] Granted, this would only be a meaningful transition for workers if they are supported in accessing those new jobs—but there is hope in that department, as there are already successful transitions taking place more broadly. Mike Weaver is just one of a growing number of farmers who have made the switch from raising animals for slaughter to growing plants, in this case, hemp.

Still, while the widespread adoption of cell-cultured meat may ultimately improve the welfare of humans, this doesn't mean we'll expand our moral circle to include nonhuman animals. As philosopher Ben Bramble puts it in *The Conversation*, "The moral problem stems from the fact that we will likely switch over to lab-grown meat because it is cheap, or thanks to its benefits for human health or the environment. That is, we will do it for our own sake and not for the sake of animals."[24] He continues, "If we switch over to lab-grown meat just for our own sake, and not for the sake of animals, then the morally dubious part of us that is responsible for our inaction on factory farming will remain intact. If this part of us has other bad consequences, then we might have lost a valuable opportunity to confront it and avoid those outcomes." In other words, this would essentially amount to what we have achieved in the past, which is a game of ethical Whack-a-Mole.

Consider that we've saved whales by creating kerosene (eliminating the need to extract oil from their blubber), spared geese by developing pens (rather than plucking their feathers for quills), helped pigs by engineering insulin in bacteria (doing away with the practice of harvesting it from their pancreases), and so on. Using various forms of technology, we ended exploitation of one group of animals in one context, only to see another form of oppression remain or pop up. So too may be the case with factory farming. This would be because we hadn't been able to build a moral ethic of not harming animals. We might surmise, for instance, that if slavery had ended purely because of technological advances, and if there had never been a Civil War motivated by moral considerations against slavery, Americans would have lost an opportunity to more fully confront its evils. Perhaps similar considerations hold true whenever any moral ill related to the exploitation of others for products is addressed solely by finding an alternative, nonexploitative means of creating the same products—involving no sacrifice or moral reckoning on anyone's part.

On the flip side, some farmed animal advocates, such as Ezra Klein, founder of *Vox* and now a columnist at the *New York Times*, believe cell-cultured meat will help facilitate moral progress. According to Klein, "Nothing changes a society's values as quickly as innovations that make a new moral system easy and cheap to adopt."[25] Many people have pointed out that giving up animal products for health or environmental reasons can lead to greater concern for animals, because once people no longer have a need to rationalize animal product consumption, they can take a harder look at the ethics of it. This is the view of Tobias Leenaert, author of *How to Create a Vegan World: A Pragmatic Approach*, who suspects cell-cultured meat will "be the technological revolution that precedes a moral revolution."[26] But this too could be seen as a moral failure, because it implies that it's only once we do not stand to benefit from a practice that we can appreciate how morally atrocious it is.

If we zoom out from current generations and instead consider how the adoption of cell-cultured meat could change eating habits and ethics across future generations, it's harder to maintain a pessimistic view. Those who grow up in a world in which there is no industrial animal farming are more likely to be growing up in a culture that takes animal lives and suffering more seriously than our own. Even if these cultural changes come about through cell-cultured meat, young people are likely to internalize these moral values and take them as a given. In other words, if their pro-animal values are firmly and sincerely accepted, it would be overly nitpicky to fret about how societies arrived at these values; what is important is that they exist.

Despite its promise, whether or not cell-cultured meat will see the light of day in supermarkets and restaurants around the globe remains to be seen. Purveyors of cell-cultured meat will first have to clear many hurdles before they can sell their products to the masses. But if cell-cultured meat does someday hit the shelves and our dinner plates, it may not take long for it to seem far more normal than eating meat from animal bodies. It will just be what it looks like: packaged protein with no relationship to the death of a living being.

Epilogue

It is precisely the minor differences in people who are otherwise alike that form the basis of feelings of hostility between them. —*Sigmund Freud*

Meat consumption is deeply entrenched in our lives for countless historical and contemporary reasons. In the earliest days, humans scavenged the remains of animals that had fallen prey to their natural predators. Today, we farm billions of animals in overcrowded factories. From its humble beginnings above our fire pits to its current place as the centerpiece of almost every meal, meat is now so cheap, convenient, and tasty that for most Americans, it's simply a given. The seductive nature of meat, and the integral role it has come to play in our religious and social norms, as well as the industries that spend billions of dollars to keep it on our plates will be difficult to overcome.

Nevertheless, amid these challenges, I see glimmers of hope. More and more people are speaking out against the ethical horrors, environmental destruction, and human health consequences of factory farming. More and more, people are interested in cutting back on their meat consumption.*

---

* To my surprise, in 2019, my dad started to cut back on animal products. It happened because he rediscovered his love for exercise, and when a trainer told him he was "throwing it all away," he made the shift. After years of me trying to persuade him to eat differently, it was someone else who convinced him—although I like to think I planted the seed. Today, he's enjoying being twenty pounds lighter—what he calls "the bowling ball" he used to carry around—and has boundless energy. He even calls me from time to time to tell me that he is enjoying a kale smoothie. If he can do it, anybody can.

More and more alternatives to industrial meat are at our disposal than ever before. While each alternative has pros and cons, all show potential to be disruptive and effective. What's more, it's likely that these approaches can and do complement one another. No one option—among plant-based meat, better meat, and cell-cultured meat—needs to be the only challenger to industrial meat; indeed, it will be easier to end factory farming when there are many ethical alternatives catering to a wide variety of taste preferences and moral outlooks. Not everyone will make room on their plate for bleeding veggie burgers, pampered pigs, or flesh cultured from cells, but there is a role and ethical case for each of these alternatives, and all are unquestionably better than industrial animal farming. Given that there are no silver bullets—and there are real consequences to not revolutionizing the way we currently produce most meat—it is critical that we work together on all of these paths to make them a widespread reality. But that won't necessarily be easy.

It turns out people whose ideologies are close but not identical often seem to dislike each other the most. Consider the quarrels among various brands of progressives (such as supporters of Bernie Sanders and supporters of Hillary Clinton) captured in the phrase "the left eats its own." Freud called this phenomenon "the narcissism of small differences." Research has referred to it as "horizontal hostility," which grows as a reaction to "distinctiveness threat." The distinctness under threat is differentiation from mainstream groups, a differentiation minority groups usually value as part of their identities. "The more narrowly focused a political identity, the harder it is to build political solidarity,"[1] Samuel Boerboom, editor of *The Political Language of Food*, explained to me.

As Adam Grant described in his bestselling book *Originals: How Non-Conformists Move the World*, psychologists have studied this dynamic among strict vegans and less-strict vegetarians, finding some vegans were more hostile toward vegetarians than they were toward omnivores. The researchers assumed this was about identity protection—when vegans, vegetarians, and flexitarians (who avoid meat and dairy but aren't absolutists) are lumped together, the vegan identity gets diluted by association.

Horizontal hostility across this movement makes sense psychologically, but it's a strategic disaster. It leads to infighting among people with overlapping goals, which makes it counterproductive. As historian James

McWilliams wrote, "As has so often been the case with reform movements, infighting between those who seek evolution and those who seek revolution fosters more stagnation than progress."[2]

Ultimately, we all want to see the end of factory farming. Yet we waste time by focusing on things like virtue points and identity shoring, and disputing each other's visions of the ideal post-factory-farming landscape.

What we should prioritize instead is the end of factory farming. To do that, we must support, or at least not actively oppose, legal approaches toward that end—even when others' solutions for chipping away at factory farming are not our preferred ones. This means plant-based meat and cell-cultured meat advocates not actively opposing better meat—even if they don't think better meat is the ethical, environmental, or nutritional ideal. This also of course means better-meat advocates not opposing plant-based and cell-cultured meat for not being "the real thing." No one fighting for change through any of these fronts should disagree that it's factory farming that has to go—not any of the alternatives seeking to displace it. Plant-based meat, better meat, and cell-cultured meat each chip away at industrial animal farming in a way the others can't. For now, at least, we need them all.

One thing that stood out to me as I learned more about how industrial animal agriculture took over the world was the dominant role of profit. While the innovators who turned animal agriculture into a flesh-generating empire may have paid lip service to positive motivations along the way—think Philip Armour's yellow train cars painted with the slogan, "We Feed the World"—much of that ingenuity was motivated by profit. Every step "forward" was about cutting costs and increasing output and making more money. It might once have been possible to see something admirable about this. Even though the titans of industrial agriculture were motivated by self-interest, some of them may also have felt they were improving the world by bringing cheap protein to the masses. Now we have a better idea of the human, animal, and environmental costs of cheap meat, which far outweigh any good that comes from it. Any real solution to industrial farming cannot be motivated solely by profit; there must be a desire to do good as well. It is precisely this virtue that will unite proponents of plant-based, better meat, and cell-cultured meat.

Will Harris of White Oak Pastures was doing very well as an industrial farmer, yet he changed his successful business model because he ultimately cared more about the welfare of his animals than his bottom line. Likewise, when Beyond Meat founder Ethan Brown envisioned his plant-based McDonald's, and then a company that would produce plant foods that taste like meat, it wasn't because he thought this was the best way to make money. Rather, he was motivated to improve animal welfare, the environment, and human health—but his movement caught fire and he created a multibillion-dollar company in the process. Finally, while cell-cultured meat advocates believe that creating real meat without animals will be more efficient than raising animals for food—and thus eventually more profitable—that's far from the case now. If all these pioneers were concerned solely with money, they would not be devoting themselves to a technology that may take decades to turn a profit. Many of them are moving ahead despite such extreme scalability difficulties because they're also motivated by a concern for human health, animal welfare, and the environment. Indeed, many of today's cell-cultured meat companies were founded and backed by animal advocates.

As we support these alternatives to factory farming, we need to be aware of how polarized this debate will become as it grows more enmeshed with politics and the fight over climate change—all of which is extremely polarized in the United States and in much of the world right now. Those who see averting climate change as part of a nefarious left-wing agenda to control people, slow our economies, and drain pleasure from our lives will be suspicious of alternatives to industrial meat that are advertised as more climate friendly. This makes it all the more important that the coalition against factory farming be a broad one, tactically, culturally, and philosophically.

We will never fully agree on principles and tactics. At the same time, we'll always have overlapping concerns, and those should be our central focus. To the extent all of us can work together, we should. When the differences are too large to make that possible, we can take separate paths. But what we can never do is turn on each other. There is too much at stake—and I think all of us can agree on that.

Notes

INTRODUCTION

1. Jacy Reese, "There's No Such Thing as Humane Meat or Eggs. Stop Kidding Yourself," *Guardian*, Nov. 16, 2018, https://www.theguardian.com/food/2018/nov/16/theres-no-such-thing-as-humane-meat-or-eggs-stop-kidding-yourself.

CHAPTER 1

1. 99% Invisible, "What Was the First Item Ever Designed?," *Slate*, Sep. 12, 2014, http://www.slate.com/blogs/the_eye/2014/09/12/_99_invisible_roman_mars_on_the_acheulean_hand_ax_and_the_genesis_objects.html.

2. Lynne Rossetto Kasper, "Why Do We Eat Meat? Tracing the Evolutionary History," *The Splendid Table*, March 6, 2013, https://www.splendidtable.org/story/2013/03/06/why-do-we-eat-meat-tracing-the-evolutionary-history.

3. Jeffrey Kluger, "Sorry Vegans: Here's How Meat-Eating Made Us Human," *Time*, March 9, 2016, http://time.com/4252373/meat-eating-veganism-evolution/.

4. Katherine Zink and Daniel Lieberman, "Impact of Meat and Lower Palaeolithic Food Processing Techniques on Chewing in Humans," *Nature* 531, (March 9, 2016): 500–503.

5. Richard Wrangham, *Catching Fire: How Cooking Made Us Human* (New York: Basic Books, 2009), 8.

6. Interview with Chris Stringer, Oct. 13, 2016.

CHAPTER 3

1. "Christopher Columbus," History.com, updated Oct. 7, 2020, https://www.history.com/topics/exploration/christopher-columbus.

2. Clements R. Markham (ed.), *The Journal of Christopher Columbus (During His First Voyage, 1492–93)* (London: Elibron Classics, 2005), 54.

3. "History of Pork," Texas Pork Producers Association, https://texaspork.org/history-of-pork/.

4. Rebecca Earle, "'If You Eat Their Food . . .': Diets and Bodies in Early Colonial Spanish America," *American Historical Review* 115, no. 3 (June 2010): 688–713.

5. Susanne Groom, *At the King's Table: Royal Dining through the Ages* (London: Merrell, 2013), eBook.

6. Jeffrey L. Forgeng, *Daily Life in Stuart England* (Westport, CT: Greenwood Press, 2007), 153.

7. Virginia DeJohn Anderson, *Creatures of Empire: How Domestic Animals Transformed Early America* (New York: Oxford University Press, 2004), eBook.

8. Andrew F. Smith, *Food in America: The Past, Present, and Future of Food, Farming, and the Family Meal* (Santa Barbara, CA: ABC-CLIO, 2017), 221.

9. Katharine E. Harbury, *Colonial Virginia's Cooking Dynasty* (Columbia: University of South Carolina Press, 2004), 75.

10. John Lawson, *A New Voyage to Carolina* (Chapel Hill: University of North Carolina Press, 2000), eBook.

11. Clayton Colman Hall (ed.), *Narratives of Early Maryland, 1633–1684* (New York: Charles Scribner's Sons, 1910), 291.

12. Daniel Blake Smith and Lorri Glover, *The Shipwreck That Saved Jamestown: The Sea Venture Castaways and the Fate of America* (New York: Henry Holt, 2008), 38.

13. Nicholas Hayward, *A Huguenot Exile in Virginia: Or, Voyages of a Frenchman Exiled for His Religion, with a Description of Virginia & Maryland* (New York: Press of the Pioneers, 1934), 123.

14. Sarah F. McMahon, "A Comfortable Subsistence: The Changing Composition of Diet in Rural New England, 1620–1840," *William and Mary Quarterly* 42, no. 4 (April 1985).

15. Stanley Lebergott, *Pursuing Happiness: American Consumers in the Twentieth Century* (Princeton, NJ: Princeton University Press, 2014), 79.

16. Abigail Carroll, *Three Squares: The Invention of the American Meal* (New York: Basic Books, 2013), 30.

17. Amelia Simmons, *American Cookery* (Carlisle, MA: Applewood Books, 1996), eBook.

18. Lydia Maria Child, *The American Frugal Housewife: Dedicated to Those Who Are Not Ashamed of Economy* (Carlisle, MA: Applewood Books, 1989), 43.

19. Jerry Robinson, *Bankruptcy of Our Nation (Revised and Expanded): Your Financial Survival Guide* (Green Forest, AR: New Leaf Publishing Group, 2012), 99–100.

CHAPTER 4

1. Isabella Lucy Bird, *The Englishwoman in America* (London: John Murray, 1856), 125.

2. Brent Coleman, "They Didn't Fly, but Pigs Once Roamed Cincinnati Streets by the Thousands in a Meat-Packing Marathon," WCPO, May 2, 2017, https://www.wcpo.com/news/insider/they-didnt-fly-but-pigs-once-roamed-cincinnati-streets-by-the-thousands-in-a-meat-packing-marathon.

3. *Report of the Secretary of Agriculture* (Washington, D.C.: US Government Printing Office, 1867), 386.

4. "The New Porkopolis," *New York Times*, March 28, 1863, https://www.nytimes.com/1863/03/28/archives/the-new-porkopoliscincinnati-has.html?searchResultPosition=1.

5. William Cronon, *Nature's Metropolis: Chicago and the Great West* (New York: W.W. Norton, 1992), 349.

6. Milo Milton Quaife (ed.), *The Development of Chicago, 1674–1914: Shown in a Series of Contemporary Original Narratives* (Chicago: Caxton Club, 1916), 239–41.

7. David Riesman, *The American City: A Sourcebook of Urban Imagery* (New York: Routledge, 2017), 45.

8. Jennifer Jensen Wallach, *How America Eats: A Social History of U.S. Food and Culture* (Lanham, MD: Rowman & Littlefield, 2013), 50.

9. J. Frank Dobie, *The Longhorns* (Austin: University of Texas Press, 1980), 147.

10. Charles A. Siringo, *A Texas Cow-Boy* (Chicago: M. Umbdenstock, 1885), 123.

11. *Chicago: A Strangers' and Tourists' Guide to the City of Chicago* (Chicago: Relig. Philo. Pub. Association [J. S. Thompson], 1866), 59.

12. Jonathan Rees, *Before the Refrigerator: How We Used to Get Ice* (Baltimore: Johns Hopkins University Press, 2018), 53.

13. Sydney Anderson, "Producing Live Stock for the Meat Industry," *Butcher's Advocate and Market Journal*, July 26, 1922.

14. *American Railroad Journal and Mechanics' Magazine*, 1842, 165.

15. Michael North, *Machine-Age Comedy* (New York: Oxford University Press, 2009), 10.

16. Ted Genoways, *The Chain: Farm, Factory, and the Fate of Our Food* (New York: HarperCollins, 2015), eBook.

CHAPTER 5

1. Interview with Anita Krajnc, July 1, 2018.

2. Andrew F. Smith, *Food in America: The Past, Present, and Future of Food, Farming, and the Family Meal* (Santa Barbara, CA: ABC-CLIO, 2017), 40.

3. Ibid.

4. Minnesota Farmers' Institute, *Annual No. 7* (Minneapolis: O.C. Gregg, Sup't, 1894), 192.

5. William Henry Williams, *Delmarva's Chicken Industry: 75 Years of Progress* (Georgetown, DE: Delmarva Poultry Industry, Inc., 1998).

6. Emelyn Rude, *Tastes Like Chicken: A History of America's Favorite Bird* (New York: Pegasus Books, 2016), eBook.

7. Maureen Ogle, *In Meat We Trust: An Unexpected History of Carnivore America* (New York: Houghton Mifflin Harcourt, 2013), 90.

8. John Fraser Hart, *The Changing Scale of American Agriculture* (Charlottesville: University of Virginia Press, 2003), 117.

9. Ibid., 116.

10. National Poultry Improvement Plan, "Poultry Disease Information," U.S. Poultry, http://www.poultryimprovement.org/default.cfm?CFID=588848&CFTOKE N=eddf5ec53f4e552c-7641D891-9C56-C7B0-AE759E6AC8C0E4D8.

11. J. C. Walker, "Benjamin Minge Duggar," in *Biographical Memoirs* (Washington, D.C.: National Academy of Sciences, 1877), 118–19, http://www.nasonline.org/ publications/biographical-memoirs/memoir-pdfs/duggar-benjamin.pdf.

12. Richard Conniff, "The Man Who Turned Antibiotics into Animal Feed—Part 3," *Strange Behaviors* (blog), March 10, 2014, https://strangebehaviors.wordpress .com/2014/03/10/the-man-who-turned-antibiotics-into-animal-feed-part-3/.

13. Thomas Jukes and E. L. Robert Stokstad, "The Multiple Nature of the Animal Protein Factor," *Journal of Biological Chemistry* 180, no. 2 (Sep. 1949): 647–54.

14. Maryn McKenna, *Big Chicken: The Incredible Story of How Antibiotics Created Modern Agriculture and Changed the Way the World Eats* (Washington, D.C.: National Geographic Society, 2017), 43.

15. *Everybody's Poultry Magazine* 50 (1945), 7.

16. Rude, *Tastes Like Chicken.*

17. Alice Roberts, *Tamed: Ten Species That Changed Our World* (New York: Random House, 2017), eBook.

18. Bosley Crowther, "Celeste Holm and Dan Dailey Star in 'Chicken Every Sunday,' New Bill at the Roxy," *New York Times*, Jan. 19, 1949, https://www.nytimes .com/1949/01/19/archives/celeste-holm-and-dan-dailey-star-in-chicken-every -sunday-new-bill.html.

CHAPTER 6

1. Bob Ortega, *In Sam We Trust: The Untold Story of Sam Walton and How Wal-Mart Is Devouring the World* (London: Kogan Page, 1999), 43.

2. Ai Hisano, "Cellophane, the New Visuality, and the Creation of Self-Service Food Retailing," Harvard Business School General Management Unit Working Paper No. 17–106, May 24, 2017, https://ssrn.com/abstract=2973544.

3. Interview with Anastacia Marx De Salcedo, May 20, 2017.

4. Joe Levit, "How Do They Make Spam?," *Live Science*, Sep. 16, 2010, https://www.livescience.com/32813-hormel-spam-no-mystery-meat.html.

5. Gregory Benford et al., "The Future That Never Was: Pictures from the Past," *Popular Mechanics*, Jan. 27, 2011, https://www.popularmechanics.com/flight/g462/future-that-never-was-next-gen-tech-concepts/.

6. "TV Dinners Seek Gourmet Market," *New York Times*, Feb. 10, 1984, https://www.nytimes.com/1984/02/10/business/tv-dinners-seek-gourmet-market.html.

7. Josh Ozersky, *The Hamburger: A History* (New Haven, CT: Yale University Press, 2008), 14.

8. Ray Kroc, with Robert Anderson, *Grinding It Out: The Making of McDonald's* (New York: St. Martin's, 2016), 65.

9. Ibid., 109.

10. Greg Critser, *Fat Land: How Americans Became the Fattest People in the World* (New York: Houghton Mifflin Harcourt, 2004), 113.

11. K. Annabelle Smith, "The Fishy History of the McDonald's Filet-O-Fish Sandwich," *Smithsonian Magazine*, March 1, 2013, https://www.smithsonianmag.com/arts-culture/the-fishy-history-of-the-mcdonalds-filet-o-fish-sandwich-2912/.

12. Michael Pollan, *The Omnivore's Dilemma* (New York: Penguin, 2006), 105.

13. Drew Desilver, "Chart of the Week: Is Food Too Cheap for Our Own Good?," Pew Research, May 23, 2014, https://www.pewresearch.org/fact-tank/2014/05/23/chart-of-the-week-is-food-too-cheap-for-our-own-good/.

14. "Panel Stands by Its Dietary Goals but Eases a View on Eating Meat," *New York Times*, Jan. 24, 1978, https://www.nytimes.com/1978/01/24/archives/panel-stands-by-its-dietary-goals-but-eases-a-view-on-eating-meat.html.

CHAPTER 7

1. Mahatma Gandhi, *An Autobiography or The Story of My Experiments with Truth* (New Haven, CT: Yale University Press, 2018), 131.

2. David A. Kessler, *The End of Overeating: Taking Control of the Insatiable American Appetite* (New York: Rodale Books, 2010), 12.

3. Michael Moss, "The Extraordinary Science of Addictive Junk Food," *New York Times*, Feb. 20, 2013, https://www.nytimes.com/2013/02/24/magazine/the-extraordinary-science-of-junk-food.html.

4. Barb Stuckey, *Taste What You're Missing: The Passionate Eater's Guide to Why Good Food Tastes Good* (New York: Atria Books, 2012), 29.

5. Mark Schatzker, *The Dorito Effect: The Surprising New Truth About Food and Flavor* (New York: Simon & Schuster, 2016), 50.

6. Melanie Warner, *Pandora's Lunchbox: How Processed Food Took Over the American Meal* (New York: Scribner, 2013), 178.

7. Ben Tinker, "How Your Food Is Engineered to Taste Great," CNN, Nov. 9, 2017, https://www.cnn.com/2017/11/09/health/food-flavor-natural-artificial/index.html.

8. John Prescott, "Meet the New Taste . . . Same as the Old Taste?," LinkedIn, Feb. 17, 2015, https://www.linkedin.com/pulse/meet-new-taste-same-old-john-prescott.

9. Schatzker, *The Dorito Effect*, 52.

10. "Chicken McNuggets," McDonald's, https://www.mcdonalds.com/us/en-us/product/chicken-mcnuggets-4-piece.html.

11. Interview with Bee Wilson, Oct. 12, 2016.

12. Bee Wilson, *First Bite: How We Learn to Eat* (New York: Basic Books, 2015), 92.

CHAPTER 8

1. "There's an Answer to Cattle's Carbon Emissions—and It Isn't Less Beef," National Cattlemen's Beef Association Sponsored Article, *Quartz*, Aug. 21, 2018, https://qz.com/1363873/theres-an-answer-to-cattles-carbon-emissions-and-it-isnt-less-beef/.

2. Alexandra Bruell, "Beef Is Back for Dinner as Marketers Woo Nostalgic Millennials," *Wall Street Journal*, Oct. 5, 2017, https://www.wsj.com/articles/beef-industry-aims-to-herd-millennials-with-nostalgic-ad-1507201382.

3. Kevin Schulz, "Report: Pork Checkoff Pays Off for Pork Producers," *National Hog Farmer*, Jan. 8, 2018, https://www.nationalhogfarmer.com/business/report-pork-checkoff-pays-pork-producers.

4. Rhonda Perry, "Anti–Check Off Signatures Are Flooding In," *In Motion Magazine*, Feb. 7, 1999, http://www.inmotionmagazine.com/checko.html.

5. Chris Petersen, "Congress Must Reform Commodity Checkoff Programs," *Des Moines Register*, June 18, 2018, https://eu.desmoinesregister.com/story/opinion/columnists/iowa-view/2018/06/18/new-farm-bill-reform-misused-commodity-checkoff-programs-factory-farming/705365002/.

6. *New York Times* Editorial Board, "The Other Political Pork," *New York Times*, Nov. 10, 2002, https://www.nytimes.com/2002/11/10/opinion/the-other-political-pork.html.

7. "Paid Advertisement: Beefing Up Sustainability," *Wall Street Journal*, Aug. 14, 2021, A5.

8. Deena Shanker, "The US Meat Industry's Wildly Successful, 40-Year Crusade to Keep Its Hold on the American Diet," *Quartz*, updated Oct. 22, 2015, https://qz.com/523255/the-us-meat-industrys-wildly-successful-40-year-crusade-to-keep-its-hold-on-the-american-diet/.

9. Charlie Mitchell and Austin Frerick, "The Hog Barons," *Vox*, updated April 19, 2021, https://www.vox.com/the-highlight/22344953/iowa-select-jeff-hansen-pork-farming.

10. Steve Johnson, "The Politics of Meat," PBS *Frontline*, https://www.pbs.org/wgbh/pages/frontline/shows/meat/politics/.

11. Charles Siderius, "Off the Killing Floor," *Dallas Observer*, Jan. 11, 2001, https://www.dallasobserver.com/news/off-the-killing-floor-6393233.

12. "How the Heart-Check Food Certification Program Works," American Heart Association, https://www.heart.org/en/healthy-living/company-collaboration/heart-check-certification/how-the-heart-check-food-certification-program-works.

13. "Lobbyists Want to Reverse Michelle Obama's Healthy School Lunch Program," *The Grio,* March 15, 2017, https://thegrio.com/2017/03/15/lobbyists-want-to-reverse-michelle-obamas-healthy-school-lunch-program/.

14. Allison Aubrey, "More Salt, Fewer Whole Grains: USDA Eases School Lunch Nutrition Rules," NPR, *The Salt*, Dec. 7, 2018, https://www.npr.org/sections/thesalt/2018/12/07/674533555/more-salt-in-school-lunch-fewer-whole-grains-usda-eases-school-lunch-rules.

15. Sam Howe Verhovek, "Talk of the Town: Burgers v. Oprah," *New York Times*, Jan. 21, 1998, https://www.nytimes.com/1998/01/21/us/talk-of-the-town-burgers-v -oprah.html.

16. Associated Press, "Dip in Beef Prices Seasonal, Expert Says in Oprah Trial, *Los Angeles Times*, Feb. 12, 1998, https://www.latimes.com/archives/la-xpm-1998-feb-12 -mn-18386-story.html.

17. Sue Anne Pressley, "Oprah Winfrey Wins Case Filed by Cattlemen," *Washington Post*, Feb. 27, 1998, https://www.washingtonpost.com/archive/ politics/1998/02/27/oprah-winfrey-wins-case-filed-by-cattlemen/dd4612f5-ccbf -4e3d-a1c1-f84d1f4fd23c/.

18. "What PETA REALLY Stands For," PETA, https://www.peta.org/features/ what-peta-really-stands-for/; https://www.peta.org/issues/animals-used-for -experimentation/silver-spring-monkeys/.

19. Interview with Daisy Freund, Nov. 14, 2018.

20. "What Is Ag-Gag Legislation?," American Society for the Prevention of Cruelty to Animals, https://www.aspca.org/improving-laws-animals/public-policy/what-ag -gag-legislation.

21. Richard Oppel Jr., "Taping of a Farm Cruelty Is Becoming the Crime," *New York Times*, April 6, 2013, https://www.nytimes.com/2013/04/07/us/taping-of-farm -cruelty-is-becoming-the-crime.html.

22. "What Is Ag-Gag Legislation?" ASPCA, https://www.aspca.org/improving-laws- animals/public-policy/what-ag-gag-legislation

23. Ron Nixon, "Farm Bill Compromise Will Change Programs and Reduce Spending," *New York Times*, Jan. 27, 2014, https://www.nytimes.com/2014/01/28/ us/politics/farm-bill-compromise-will-reduce-spending-and-change-programs.html.

24. Robert Paarlberg, *Food Politics: What Everyone Needs to Know* (New York: Oxford University Press, 2013).

25. Dan Charles, "Farmers Got Billions from Taxpayers in 2019, and Hardly Anyone Objected," NPR, *All Things Considered*, Dec. 31, 2019, https://www.npr.org/ sections/thesalt/2019/12/31/790261705/farmers-got-billions-from-taxpayers-in-2019 -and-hardly-anyone-objected.

26. Geoff Dembicki, "Trump Is Bailing Out Big Meat—and Further Screwing the Planet," *New Republic*, June 1, 2020, https://newrepublic.com/article/157913/trump -bailing-big-meatand-screwing-planet.

27. Dan Imhoff, *Food Fight: The Citizen's Guide to the Next Food and Farm Bill* (Plymouth, MA: Watershed Media, 2012), eBook.

28. Jonathan Barker, "Poverty, Hunger, and US Agricultural Policy: Do Farm Programs Affect the Nutrition of Poor Americans?," American Enterprise Institute, Jan. 27, 2017, https://aic.ucdavis.edu/2017/01/27/poverty-hunger-and-us -agricultural-policy-do-farm-programs-affect-the-nutrition-of-poor-americans/.

29. Jayson Lusk, "Are Farm Subsidies Making Us Fat?," *Jayson Lusk: Food and Agricultural Economist* (blog), July 22, 2016, http://jaysonlusk.com/blog/2016/7/22/ are-farm-subsidies-making-us-fat.

30. Brian Kahn, "Joe Biden Is Not Coming for Your Poorly Cooked Hamburger," *Gizmodo*, April 26, 2021, https://earther.gizmodo.com/joe-biden-is-not-coming-for -your-poorly-cooked-hamburge-1846764056.

31. United States Department of Agriculture, "Dietary Guidelines for Americans 2020–2025," https://www.dietaryguidelines.gov/sites/default/files/2020-12/Dietary_ Guidelines_for_Americans_2020-2025.pdf.

32. Brian Kateman, "Pasta Is Now a Vegetable? USDA's School Lunch Guidelines Threaten the Health of Our Nation's Children," *Forbes*, Jan. 30, 2020, https://www .forbes.com/sites/briankateman/2020/01/30/pasta-is-now-a-vegetable-usdas-school -lunch-guidelines-threaten-the-health-of-our-nations-children/.

33. Marion Nestle, *Unsavory Truth: How Food Companies Skew the Science of What We Eat* (New York: Basic Books, 2018), eBook.

34. Marion Nestle, "At Last: The 2020 Dietary Guidelines Advisory Committee," *Food Politics* (blog), Feb. 26, 2019, https://www.foodpolitics.com/2019/02/at-last-the -2020-dietary-guidelines-advisory-committee/.

35. Interview with Walter Willett, Aug. 8, 2018.

36. Lauren O'Connor et al., "A Mediterranean-Style Eating Pattern with Lean, Unprocessed Red Meat Has Cardiometabolic Benefits for Adults Who Are Overweight or Obese in a Randomized, Crossover, Controlled Feeding Trial," *American Journal of Clinical Nutrition* 108, no. 1 (July 2018): 33–40.

37. Amy Patterson Neubert, "Study: DASH Diet Can Substitute Lean Pork for Chicken or Fish to Reduce Blood Pressure," Purdue University, June 10, 2015, https://www.purdue.edu/newsroom/releases/2015/Q2/study-dash-diet-can -substitute-lean-pork-for-chicken-or-fish-to-reduce-blood-pressure.html.

38. Nestle, *Unsavory Truth.*

39. Sanjiv Agarwal, "Association of Lunch Meat Consumption with Nutrient Intake, Diet Quality and Health Risk Factors in U.S. Children and Adults: NHANES 2007–2010," *Nutrition Journal* 14 (Dec. 2015): 128.

40. Interview with David Katz, Aug. 14, 2018.

41. Peter Whoriskey, "Hot Dogs, Bacon and Other Processed Meats Cause Cancer, World Health Organization Declares," *Washington Post*, Oct. 26, 2015, http://www .washingtonpost.com/news/wonkblog/wp/2015/10/26/hot-dogs-bacon-and-other -processed-meats-cause-cancer-world-health-organization-declares/.

42. "IARC Meat Vote Is Dramatic and Alarmist Overreach," North American Meat Institute, Oct. 26, 2015, https://www.meatinstitute.org/index.php?ht=display/ ReleaseDetails/i/116652/pid/287.

43. Nanci Hellmich, "Red-Meat Intake Linked to Increased Risk of Diabetes," *USA Today*, June 17, 2013, https://www.usatoday.com/story/news/nation/2013/06/17/ diabetes-red-meat-intake/2431405/.

44. Kayla James, "'Don't Give Up': Man Paralyzed in 2015 Shooting Heads to Paralympic Games," WCVB, updated June 28, 2021, https://www.wcvb.com/article/ iowan-paralyzed-2015-shooting-heads-to-paralympic-games/36859930.

45. Kathleen Doheny, "Eating Red Meat May Boost Death Risk," MedicineNet, March 23, 2009, https://www.medicinenet.com/script/main/art .asp?articlekey=98724.

46. Sigal Samuel, "It's Not Just Big Oil. Big Meat Also Spends Millions to Crush Good Climate Policy," *Vox*, April 13, 2021, https://www.vox.com/future -perfect/22379909/big-meat-companies-spend-millions-lobbying-climate.

47. Lily Rothman, "Link Between Bacon and Cancer Had Been Long Suspected," *Time*, Oct. 26, 2015, https://time.com/4086914/bacon-cancer-nitrate-1980/.

48. Judy Blitman, "Food and Health Experts Warn against Bringing Home the Bacon," *New York Times*, Aug. 8, 1973, https://www.nytimes.com/1973/08/08/ archives/food-and-health-experts-warn-against-bringing-home-the-bacon-im.html.

49. Philip Hilts, "The Day Bacon Was Declared Poison," *Washington Post*, April 26, 1981, https://www.washingtonpost.com/archive/lifestyle/magazine/1981/04/26/the-day-bacon-was-declared-poison/07f7c03e-44a9-45cd-9204-f0964096774d/.

50. Ibid.

51. Victor Cohn, "U.S. Agencies Reject Banning Nitrite in Meat," *Washington Post*, Aug. 20, 1980, https://www.washingtonpost.com/archive/politics/1980/08/20/us-agencies-reject-banning-nitrite-in-meat/d42a3a2c-77c2-4514-8996-baf8a643e251/.

52. "Beef Cattle Production in the US Not a Significant Contributor to Long-Term Global Warming," Farms.com, March 12, 2019, https://www.farms.com/news/beef-cattle-production-in-the-us-not-a-significant-contributor-to-long-term-global-warming-144140.aspx.

53. Marion Nestle, "Least Credible Food Industry Ad of the Week: JBS and Climate Change," *Food Politics* (blog), April 26, 2021, https://www.foodpolitics.com/2021/04/least-credible-food-industry-ad-of-the-week-jbs-and-climate-change/.

54. Ashoka Mukpo, "JBS, Other Brazil Meatpackers Linked to Devastating Pantanal Fires, Greenpeace Says," *Mongabay*, March 17, 2021, https://news.mongabay.com/2021/03/jbs-other-brazil-meatpackers-linked-to-devastating-pantanal-fires-greenpeace-says/.

55. Aurora Solá, "Beef Giant JBS Vows to Go Deforestation-Free—14 Years from Now," *Mongabay*, April 6, 2021, https://news.mongabay.com/2021/04/beef-giant-jbs-vows-to-go-deforestation-free-14-years-from-now/.

CHAPTER 9

1. Emily Crane, "How Biden's Climate Plan Could Limit You to Eat Just One Burger a MONTH, Cost $3.5K a Year Per Person in Taxes, Force You to Spend $55K on an Electric Car and 'Crush' American Jobs," *Daily Mail*, updated April 27, 2021, https://www.dailymail.co.uk/news/article-9501565/How-Bidens-climate-plan-affect-everyday-Americans.html.

2. Marjorie Taylor Green, Twitter post, April 25, 2021, 1:57 p.m., https://twitter.com/mtgreenee/status/1386378888465485830?lang=en.

3. Philip Bump, "Fox News's Latest Straw Man Is Made of Red Meat," *Washington Post*, April 26, 2021, https://www.washingtonpost.com/politics/2021/04/26/fox-newss-latest-straw-man-is-made-red-meat/.

4. Donald Trump Jr., Twitter post, April 24, 2021, 12:16 p.m., https://twitter.com/donaldjtrumpjr/status/1385991099299377159?lang=en.

5. Lauren Boebert, Twitter post, April 24, 2021, 9:53 a.m., https://twitter.com/laurenboebert/status/1385955098715729926?lang=en.

6. Greg Abbott, Twitter post, April 25, 2021, 5:35 p.m., https://twitter.com/gregabbott_tx/status/1386433630747070468?lang=en.

7. *Facts First* (blog), "Does Biden's Climate Plan Include 'Cutting 90% of Red Meat from Our Diets by 2030'?," CNN, https://www.cnn.com/factsfirst/politics/factcheck_e5e088b0-0b69-400b-aa5d-b5cfb9168d33.

8. Gordon Hodson, "The Meat Paradox: Loving but Exploiting Animals," *Psychology Today*, March 3, 2014, https://www.psychologytoday.com/us/blog/without-prejudice/201403/the-meat-paradox-loving-exploiting-animals.

9. Kelly Weill, "Why Right Wingers Are Going Crazy about Meat," *Daily Beast*, Aug. 25, 2018, https://www.thedailybeast.com/why-right-wingers-are-going-crazy-about-meat.

10. Ibid.

11. "Meat Processing and Products," Center for Responsive Politics, https://www.opensecrets.org/industries/indus.php?ind=G2300.

12. Ryan McCrimmon, "White House Dances around a Big Contributor to Climate Change: Agriculture," *Politico*, April 22, 2021, https://www.politico.com/news/2021/04/22/climate-change-biden-agriculture-484351.

13. Paula Goodyer, "Meat Eaters Justify Diet Using 'Four Ns': Natural, Necessary, Normal, Nice," *Sydney Morning Herald*, May 30, 2015, https://www.smh.com.au/lifestyle/health-and-wellness/meat-eaters-justify-diet-using-four-ns-natural-necessary-normal-nice-20150530-ghd5le.html.

14. Weill, "Why Right Wingers Are Going Crazy about Meat."

15. Paul Rozin et al., "Is Meat Male? A Quantitative Multimethod Framework to Establish Metaphoric Relationships," *Journal of Consumer Research* 39, no. 3 (Oct. 2012): 629–43.

16. 13redcliffe, "Kingsford Charcoal Commercial," YouTube video, 0:30, April 7, 2009, https://www.youtube.com/watch?v=FyuaCLM_syA.

17. Meghan Casserly, "Grilling, Guys and the Great Gender Divide," *Forbes*, July 1, 2010, https://www.forbes.com/2010/07/01/grilling-men-women-barbecue-forbes -woman-time-cooking.html?sh=6207f7ecbad6.

18. "Sexual Politics of Meat," CarolJAdams.com, https://caroljadams.com/spom-the -book.

19. University of Technology, Sydney, "The Link between Meat and Social Status," Phys.org, https://phys.org/news/2018-09-link-meat-social-status.html.

20. Krishna Ramanujan, "Eating Green Could Be in Your Genes," *Cornell Chronicle*, March 29, 2016, https://news.cornell.edu/stories/2016/03/eating-green -could-be-your-genes.

CHAPTER 10

1. Interview with Will Harris, Jan. 18, 2019.

2. Gosia Gozniacka, "Big Food Is Betting on Regenerative Agriculture to Thwart Climate Change," *Civil Eats*, Oct. 29, 2019, https://civileats.com/2019/10/29/big -food-is-betting-on-regenerative-agriculture-to-thwart-climate-change/.

3. Larissa Zimberoff, "There's a New 'Organic' Food That Fights Global Warming," *Bloomberg*, April 23, 2021, https://www.bloomberg.com/news/articles/2021-04-23/ regenerative-farming-is-a-new-kind-of-organic-food-that-s-good-for-earth-too.

4. Jonathan Foley, "Beef Rules," GlobalEcoGuy.org, June 19, 2019, https:// globalecoguy.org/beef-rules-d5bbf65a24e3.

5. Matthew Hayek and Rachael Garrett, "Nationwide Shift to Grass-Fed Beef Requires Larger Cattle Population," *Environmental Research Letters* 13, no. 8 (July 2018): 13.

6. "Back to Grass: The Market Potential for U.S. Grassfed Beef," Stone Barns Center for Food & Agriculture, April 2017, https://www.stonebarnscenter.org/wp-content/ uploads/2017/10/Grassfed_Full_v2.pdf.

7. Tara Garnett, Cécile Godde, et al., "Grazed and Confused?" Report by Food Climate Research Network, 2017, https://www.oxfordmartin.ox.ac.uk/downloads/ reports/fcrn_gnc_report.pdf.

8. James McWilliams, "The Myth of Sustainable Meat," *New York Times*, April 12, 2012, https://www.nytimes.com/2012/04/13/opinion/the-myth-of-sustainable-meat .html.

9. Joel Salatin, "Joel Salatin responds to *New York Times*' 'Myth of Sustainable Meat,'" *Grist,* April 17, 2012, https://grist.org/sustainable-farming/farmer-responds -to-the-new-york-times-re-sustainable-meat/.

10. David Bronner, "How the Regenerative Agriculture and Animal Welfare Movements Can End Factory Farming, Restore Soil and Mitigate Climate Change," *Dr. Bronner's All-One* (blog), March 7, 2017, https://www.drbronner.com/all-one -blog/2017/03/regenetarians-unite/.

11. Tamar Haspel, "Is Grass-Fed Beef Really Better for You, the Animal and the Planet?," *Washington Post*, Feb. 23, 2015, https://www.washingtonpost .com/lifestyle/food/is-grass-fed-beef-really-better-for-you-the-animal-and-the -planet/2015/02/23/92733524-b6d1-11e4-9423-f3d0a1ec335c_story.html.

12. George Monbiot, "Goodbye—and Good Riddance—to Livestock Farming," *Guardian,* Oct. 4, 2017, https://www.theguardian.com/commentisfree/2017/oct/04/ livestock-farming-artificial-meat-industry-animals.

13. Interview with Helen Atthowe, July 6, 2019.

14. Associated Press, "'Green Manure' Keeps These Farmers Happy," NBC News, June 21, 2008, https://www.nbcnews.com/id/wbna25242888.

15. James McWilliams, "The Omnivore's Contradiction: That Free-Range, Organic Meat Was Still an Animal Killed for Your Dinner," *Salon*, Jan. 3, 2016, https://www .salon.com/2016/01/03/the_omnivores_contradiction_that_free_range_organic_ meat_was_still_an_animal_killed_for_your_dinner/.

16. Anna Charlton and Gary Francione, "A 'Humanely' Killed Animal Is Still Killed—and That's Wrong," *Aeon*, Sep. 8, 2017, https://aeon.co/ideas/a-humanely -killed-animal-is-still-killed-and-thats-wrong.

CHAPTER 11

1. Interview with Ethan Brown, Feb. 12, 2019.

2. Interview with Parker Lee, Feb. 12, 2019.

3. Interview with Daniel Ryan, Feb. 12, 2019.

4. Interview with Julie Wushensky, Feb. 12, 2019.

5. Lizzy Gurdus, "Beyond Meat Is Soaring, but There's No Real Evidence Its Products Are Healthier Than Real Meat, Says Former US Agriculture Chief," CNBC,

May 26, 2019, https://www.cnbc.com/2019/05/24/fmr-us-agriculture-chief-on
-beyond-meats-nutritional-reality.html.

6. Lauren Wicks, "What Is the Impossible Burger—And Is It Even Healthy?,"
CookingLight, May 28, 2019, https://www.cookinglight.com/news/is-the-impossible
-burger-healthy.

7. Alicia Kennedy, "Impossible Burgers Won't Save the Environment—They're
Just a Greenwashing Trend," *In These Times*, April 22, 2020, https://inthesetimes
.com/article/corporate-fake-meat-wont-save-us-impossible-burger-beyond-meat
-greenwashing.

8. Larissa Zimberoff, *Technically Food: Inside Silicon Valley's Mission to Change
What We Eat* (New York: Abrams, 2021), eBook.

CHAPTER 12

1. Neil Stephens and Martin Ruivenkamp, "Promise and Ontological Ambiguity in
the *In vitro* Meat Imagescape: From Laboratory Myotubes to the Cultured Burger,"
Science as Culture 25, no. 3 (July 2016): 327–55.

2. Warren James Belasco, *Meals to Come: A History of the Future of Food* (Berkeley:
University of California Press, 2006), 37.

3. Kat Eschner, "Winston Churchill Imagined the Lab-Grown Hamburger,"
Smithsonian Magazine, Dec. 1, 2017, https://www.smithsonianmag.com/smart
-news/winston-churchill-imagined-lab-grown-hamburger-180967349/.

4. "Lab-Grown Meat: How 'Moo's Law' Will Drive Innovation," *Money
Week*, April 16, 2021, https://moneyweek.com/investments/commodities/soft
-commodities/603088/lab-grown-meat-new-agrarian-revolution.

5. Lora Kolodny, "Impossible Foods CEO Pat Brown Says VCs Need to Ask
Harder Scientific Questions," *TechCrunch*, May 22, 2017, https://techcrunch
.com/2017/05/22/impossible-foods-ceo-pat-brown-says-vcs-need-to-ask-harder
-scientific-questions/.

6. Oliver Morrison, "'Cultured Meat Is Fool's Gold': Environmentalists Lock
Horns over Controversial Documentary," *Food Navigator*, updated Jan. 13, 2020,
https://www.foodnavigator.com/Article/2020/01/10/Cultured-meat-is-fool-s-gold
-Environmentalists-lock-horns-over-controversial-documentary.

7. Chase Purdy, "Why We Don't Yet Know If Cell-Cultured Meat Will Actually Fight Climate Change," *Quartz*, Feb. 20, 2019, https://qz.com/1553875/is-cell -cultured-meat-environmentally-friendly/.

8. Jane Calvert, "Synthetic Biology: Constructing Nature?," *Sociological Review* 58, no. 1 (May 2010): 95–112.

9. Dana Perls, "From Lab to Fork: Critical Questions on Laboratory-Created Animal Product Alternatives," Friends of the Earth, June 2018, http://foe.org/wp -content/uploads/2018/06/From-Lab-to-Fork-1.pdf.

10. Interview with Ruth DeFries, March 21, 2017.

11. Elaine Watson, "Cargill and Other 'Food Industry Giants' Join $17m Funding Round for Clean Meat Co Memphis Meats," *Food Navigator*, updated Aug. 23, 2017, https://www.foodnavigator-usa.com/Article/2017/08/23/Cargill-joins-funding -round-for-clean-meat-co-Memphis-Meats.

12. Charlie Mitchell, "Fake Meat, Real Profits," *Baffler,* Jan. 27, 2021, https:// thebaffler.com/latest/fake-meat-real-profits-mitchell.

13. Jenny Kleeman, "What's the Point of Lab-Grown Meat When We Can Simply Eat More Vegetables?," *Guardian*, Dec. 4, 2020, https://www.theguardian.com/ commentisfree/2020/dec/04/lab-grown-meat-cultured-protein.

14. Errol Schweizer, "We Won't End Factory Farming until We Dismantle the System," *Eat for the Planet* (podcast), April 21, 2021, https://eftp.co/errol-schweizer.

15. Tom Philpott, "Is Lab Meat about to Hit Your Dinner Plate?," *Mother Jones*, Aug. 2, 2021, https://www.motherjones.com/food/2021/08/is-lab-meat-about-to-hit -your-dinner-plate/.

16. Elaine Watson, "Cultured Fish Co Finless Foods Aims to Achieve Price Parity with Bluefin Tuna by the End of 2019," *Food Navigator*, Dec. 21, 2017, https://www .foodnavigator-usa.com/Article/2017/12/21/Finless-Foods-co-founder-talks-clean -meat-clean-fish-cultured-meat#.

17. Jade Scipioni, "This Restaurant Will Be the First Ever to Serve Lab-Grown Chicken (for $23)," CNBC, *Make It*, updated Dec. 23, 2020, https://www.cnbc .com/2020/12/18/singapore-restaurant-first-ever-to-serve-eat-just-lab-grown -chicken.html.

18. Sam Bloch, "The Hype and the Hope Surrounding Lab-Grown Meat," *The Counter*, July 23, 2019, https://newfoodeconomy.org/new-harvest-cell-cultured-meat-lab-meat/.

19. Martin Rowe, "Beyond Impossible: The Futures of Plant-Based and Cellular Meat and Dairy," BrighterGreen, 2019, https://brightergreen.org/wp-content/uploads/2019/07/Beyond-the-Impossible.pdf.

20. Chase Purdy, "Meet the Mechanical Engineer Who Left Boeing to Make Plant-Based Chicken," *Quartz*, Feb. 4, 2020, https://qz.com/1794108/a-former-boeing-engineer-is-making-plant-based-meat/.

21. Emma Cosgrove, "What Do Farmers Think about Cultured Meat?," *AgFunder News*, Oct. 12, 2017, https://agfundernews.com/what-do-farmers-think-about-cultured-meat.html.

22. Interview with Molly Anderson, Aug. 15, 2018.

23. Max Elder, "Cultured Meat: A New Story for the Future of Food," in *The Routledge Handbook of Animal Ethics*, edited by Bob Fischer (New York: Taylor & Francis, 2019), eBook.

24. Ben Bramble, "Lab-Grown Meat Could Let Humanity Ignore a Serious Moral Failing," *The Conversation,* Dec. 14, 2017, https://theconversation.com/lab-grown-meat-could-let-humanity-ignore-a-serious-moral-failing-88909.

25. Ezra Klein, "Ending the Age of Animal Cruelty," *Vox*, Jan. 29, 2019, https://www.vox.com/future-perfect/2019/1/29/18197907/clean-meat-cell-plant-impossible-beyond-animal-cruelty.

26. "Tobias Leenaert: 'I Suspect Clean Meat Can Be the Technological Revolution That Precedes a Moral Revolution,'" *Vegconomist*, Jan. 15, 2019, https://vegconomist.com/interviews/tobias-leenaert-i-suspect-clean-meat-can-be-the-technological-revolution-that-precedes-a-moral-revolution/.

EPILOGUE

1. Interview with Samuel Boerboom, May 20, 2017.

2. James McWilliams, "Vegan Feud," *Slate*, Sep. 7, 2012, https://slate.com/human-interest/2012/09/hsus-vs-abolitionists-vs-the-meat-industry-why-the-infighting-should-stop.html.

Further Reading

INTRODUCTION

Sentience Institute. "99% of US Farmed Animals Live on Factory Farms, Study Shows." Press release, April 11, 2019. https://www.sentienceinstitute.org/press/us -farmed-animals-live-on-factory-farms.

CHAPTER 1

Agence France-Presse. "2.4-Million-Year-Old Tools Found in Algeria Could Upend Human Origin Story." *The Telegraph*, Nov. 30, 2018. https://www.telegraph .co.uk/news/2018/11/30/24-million-year-old-tools-found-algeria-could-upend -human-origin/.

Aiello, Leslie, and Peter Wheeler. "The Expensive-Tissue Hypothesis: The Brain and the Digestive System in Human and Primate Evolution." *Current Anthropology* 36, no. 2 (April 1995): 199–221.

Boyd, Robynne. "Do People Only Use 10 Percent of Their Brains?" *Scientific American*, Feb. 7, 2008. https://www.scientificamerican.com/article/do-people -only-use-10-percent-of-their-brains/.

California Academy of Sciences. "Oldest Evidence of Stone Tool Use and Meat-Eating among Human Ancestors Discovered: Lucy's Species Butchered Meat." ScienceDaily, Aug. 11, 2020. https://www.sciencedaily.com/releases/2010/08/ 100811135039.htm.

Clinton, Keely. "Average Cranium/Brain Size of *Homo neanderthalensis* vs. *Homo sapiens*." Cobb Research Lab at Howard University, Dec. 24, 2015. https://www.cobbresearchlab.com/issue-2-1/2015/12/24/average-cranium-brain-size-of-homo-neanderthalensis-vs-homo-sapiens.

Dunn, Rob. "How to Eat Like a Chimpanzee." *Scientific American*, Aug. 2, 2012. https://blogs.scientificamerican.com/guest-blog/how-to-eat-like-a-chimpanzee/.

Joyce, Christopher. "Food for Thought: Meat-Based Diet Made Us Smarter." NPR, Aug. 2, 2010. https://www.npr.org/2010/08/02/128849908/food-for-thought-meat-based-diet-made-us-smarter.

Kluger, Jeffrey. "Sorry Vegans: Here's How Meat-Eating Made Us Human." *Time*, March 9, 2016. https://time.com/4252373/meat-eating-veganism-evolution/.

Lewin, Roger. "Man the Scavenger." *Science* 224, no. 4641 (May 1984): 861–62.

McLerran, Dan. "Study Lends New Support to Theory That Early Humans Were Scavengers." *Popular Archaeology*, March 3, 2015. https://popular-archaeology.com/article/study-lends-new-support-to-theory-that-early-humans-were-scavengers/.

Milks, Annemieke, David Parker, and Matt Pope. "External Ballistics of Pleistocene Hand-Thrown Spears: Experimental Performance Data and Implications for Human Evolution." *Scientific Reports* 9, no. 820 (Jan. 2019). https://doi.org/10.1038/s41598-018-37904-w.

Nova. "Who's Who in Human Evolution." PBS. https://www.pbs.org/wgbh/nova/hobbit/tree-nf.html.

Potts, Richard, and Christopher Sloan. *What Does It Mean to Be Human?* Washington, D.C.: National Geographic, 2010.

Semaw, Sileshi. "The World's Oldest Stone Artefacts from Gona, Ethiopia: Their Implications for Understanding Stone Technology and Patterns of Human Evolution between 2.6–1.5 Million Years Ago." *Journal of Archaeological Science* 27, no. 12 (Dec. 2000): 1197–214.

Smithsonian National Museum of Natural History. "*Ardipithecus ramidus*." https://humanorigins.si.edu/evidence/human-fossils/species/ardipithecus-ramidus.

———. "*Orrorin tugenensis*." https://humanorigins.si.edu/evidence/human-fossils/species/orrorin-tugenensis.

Wong, Kate. "How Scientists Discovered the Staggering Complexity of Human Evolution." *Scientific American*, Sep. 1, 2020. https://www.scientificamerican .com/article/how-scientists-discovered-the-staggering-complexity-of-human -evolution/.

CHAPTER 2

BBC News. "Ancient Egyptian Priests 'Killed by Rich Ritual Food.'" Feb. 26, 2010. http://news.bbc.co.uk/2/hi/health/8536480.stm.

Bekoff, Marc. "The First Domestication: How Wolves and Humans Coevolved." *Psychology Today*, Dec. 11, 2017. https://www.psychologytoday.com/us/blog/ animal-emotions/201712/the-first-domestication-how-wolves-and-humans -coevolved.

Driscoll, Carlos, et al. "From Wild Animals to Domestic Pets, an Evolutionary View of Domestication." *Proceedings of the National Academy of Sciences of the United States of America* 106, Supplement 1 (June 2009): 9971–78.

Jarus, Owen. "Who Built the Egyptian Pyramids?" *Live Science*, May 15, 2021. https://www.livescience.com/who-built-egypt-pyramids.html.

Larson, Greger. "Genetics and Domestication." *Current Anthropology* 52, no. S4 (Oct. 2011): S485–95.

Miller, Naomi, and Wilma Wetterstorm. "The Beginnings of Agriculture." In *The Cambridge World History of Food*, edited by Kenneth Kiple, 2:1123–39. London: Cambridge University Press, 2000.

Paoletta, Rae. "How Cats Conquered Humans Thousands of Years Ago." *Gizmodo*, June 19, 2017. https://gizmodo.com/how-cats-conquered-humans-thousands-of -years-ago-1796222203.

Pierotti, Raymond, and Brandy R. Fogg. *The First Domestication: How Wolves and Humans Coevolved*. New Haven, CT: Yale University Press, 2017.

United States Census Bureau. "Historical Estimates of World Population." https:// www.census.gov/data/tables/time-series/demo/international-programs/historical -est-worldpop.html.

CHAPTER 3

Anderson, James Maxwell. *The History of Portugal.* Westport, CT: Greenwood Press, 2000.

Anderson, Virginia DeJohn. *Creatures of Empire: How Domestic Animals Transformed Early America.* New York: Oxford University Press, 2004.

Bushman, Richard L. *The American Farmer in the Eighteenth Century: A Social and Cultural History.* New Haven, CT: Yale University Press, 2018.

Chandonnet, Ann. *Colonial Food.* London: Bloomsbury Publishing, 2013.

Childs, Craig. *Atlas of a Lost World: Travels in Ice Age America.* New York: Knopf Doubleday Publishing Group, 2019.

Clemen, Rudolf Alexander. *The American Livestock and Meat Industry.* New York: Ronald Press Company, 1923.

Earle, Rebecca. *The Body of the Conquistador: Food, Race and the Colonial Experience in Spanish America, 1492–1700.* New York: Cambridge University Press, 2014.

Fitzgerald, Amy J. "A Social History of the Slaughterhouse: From Inception to Contemporary Implications." *Human Ecology Review* 17, no. 1 (2010): 58–69.

Gibson, Abraham. *Feral Animals in the American South: An Evolutionary History.* New York: Cambridge University Press, 2016.

Harbury, Katharine E. *Colonial Virginia's Cooking Dynasty.* Columbia: University of South Carolina Press, 2004.

McMahon, Sarah F. "A Comfortable Subsistence: The Changing Composition of Diet in Rural New England, 1620–1840." *The William and Mary Quarterly* 42, no. 1 (1985): 26–65.

McWilliams, James E. *A Revolution in Eating: How the Quest for Food Shaped America.* New York: Columbia University Press, 2005.

Robison, Jim. *Kissimmee: Gateway to the Kissimmee River Valley.* Charleston, SC: Arcadia, 2003.

Shepherd, James F., and Samuel H. Williamson. "The Coastal Trade of the British North American Colonies, 1768–1772." *Journal of Economic History* 32, no. 4 (Dec. 1972): 783–810.

Sim, Alison. *Food and Feast in Tudor England*. Cheltenham, UK: History Press, 2005.

Stahl, Peter. "Animal Domestication in South America." In *Handbook of South American Archaeology*, edited by Helaine Silverman and William Harris Isbell, 121–30. New York: Springer, 2008.

Struzinski, Steven. "The Tavern in Colonial America." *Gettysburg Historical Journal* 1, no. 7 (2003). https://cupola.gettysburg.edu/ghj/vol1/iss1/7.

Wright, Mike. *What They Didn't Teach You about the American Revolution*. Novato, CA: Presidio Press, 2009.

CHAPTER 4

Anderson, Oscar Edward. *Refrigeration in America*. Princeton, NJ: Princeton University Press, 2015.

Anschutz, Philip F., William J. Convery, and Thomas J. Noel. *Out Where the West Begins: Profiles, Visions, and Strategies of Early Western Business Leaders*. Denver, CO: Cloud Camp Press, 2017.

Armour, Jonathan Ogden. *The Packers, the Private Car Lines, and the People*. Philadelphia: H. Altemus Company, 1906.

Avey, Tori. "Discover the History of Meatless Mondays." PBS, *The History Kitchen* (blog), Aug. 16, 2013. https://www.pbs.org/food/the-history-kitchen/history -meatless-mondays/.

Barker, Lesley. *St. Louis Gateway Rail: The 1970s*. Charleston, SC: Arcadia, 2006.

Beattie, Alan. *False Economy: A Surprising Economic History of the World*. New York: Riverhead Books, 2009.

Bigott, Joseph C. *From Cottage to Bungalow: Houses and the Working Class in Metropolitan Chicago, 1869–1929*. Chicago: University of Chicago Press, 2001.

Bjornlund, Lydia. *How the Refrigerator Changed History*. Minneapolis, MN: Abdo Publishing, 2015.

Boorstin, Daniel J. *The Americans: The Democratic Experience*. New York: Knopf Doubleday Publishing Group, 2010.

Bramley, Anne. "How Chicago's Slaughterhouse Spectacles Paved the Way for Big Meat." NPR, *The Salt*, Dec. 3, 2015. https://www.npr.org/sections/

thesalt/2015/12/03/458314767/how-chicago-s-slaughterhouse-spectacles-paved
-the-way-for-big-meat.

Brooks, Eugene Clyde. *The Story of Corn and the Westward Migration*. Chicago:
Rand McNally & Co., 1916.

Chicago Association of Commerce. "How We Began in Chicago." *Chicago
Commerce*, Sep. 24, 1921.

Clampitt, Cynthia. *Pigs, Pork, and Heartland Hogs: From Wild Boar to Baconfest*.
Lanham, MD: Rowman & Littlefield Publishers, 2018.

Coleman, Brent. "They Didn't Fly, but Pigs Once Roamed Cincinnati Streets by the
Thousands in a Meat-Packing Marathon." WCPO, May 2, 2017. https://www
.wcpo.com/news/insider/they-didnt-fly-but-pigs-once-roamed-cincinnati-streets
-by-the-thousands-in-a-meat-packing-marathon.

Cox, James. *Historical and Biographical Record of the Cattle Industry and the
Cattlemen of Texas and Adjacent Territory*. St. Louis: Woodward & Tiernan
Printing Company, 1895.

D'Eramo, Marco. *The Pig and the Skyscraper: Chicago: A History of Our Future*. New
York: Verso, 2002.

Erickson, Hal. *Any Resemblance to Actual Persons: The Real People Behind 400+
Fictional Movie Characters*. Jefferson, NC: McFarland, 2017.

Fields, Gary. *Territories of Profit: Communications, Capitalist Development, and the
Innovative Enterprises of G.F. Swift and Dell Computer*. Stanford, CA: Stanford
University Press, 2004.

Hallas, Herbert C. *William Almon Wheeler: Political Star of the North Country*.
Albany, NY: Excelsior Editions, 2013.

Hammond, Roland. *A History and Genealogy of the Descendants of William
Hammond of London, England, and His Wife Elizabeth Penn: Through Their Son
Benjamin of Sandwich and Rochester, Mass., 1600–1894*. Boston: D. Clapp & Son,
Printers, 1894.

Homans, James E., James Grant Wilson, and John Fiske (eds.). *The Cyclopædia of
American Biography*. New York: Press Association Compilers, Inc., 1915.

Horowitz, Roger. *Putting Meat on the American Table: Taste, Technology,
Transformation*. Baltimore: Johns Hopkins University Press, 2006.

Hutson, Cecil Kirk. "Texas Fever in Kansas, 1866–1930." *Agricultural History* 68, no. 1 (1994): 74–104.

Janega, James. "Meatpacking." *Chicago Tribune*, Aug. 8, 2014. https://www .chicagotribune.com/business/blue-sky/chi-meatpacking-top-chicago-innovations -bsi-series-story.html.

Kemp, Bill. "'Porkers' Wail of Anguish' Once Heard on Bloomington's South Side." *The Pantagraph*, Feb. 23, 2008. https://www.pantagraph.com/news/porkers-wail -of-anguish-once-heard-on-bloomingtons-south-side/article_dbc8dd0b-4f42- 57f6-80fd-0d157cd65224.html.

Knowlton, Christopher. *Cattle Kingdom: The Hidden History of the Cowboy West.* New York: Houghton Mifflin Harcourt, 2017.

Koziarz, Jay. "Transportation That Built Chicago: The River System." *Curbed*, Sep. 19, 2017. https://chicago.curbed.com/2017/9/19/16332590/transportation-chicago -river-history-future.

Kujovich, Mary Yeager. "The Refrigerator Car and the Growth of the American Dressed Beef Industry." *Business History Review* 44, no. 4 (1970): 460–82.

Malone, John Williams. *An Album of the American Cowboy.* New York: Watts, 1971.

McCoy, Joseph Geiting. *Historic Sketches of the Cattle Trade of the West and Southwest.* Kansas City, MO: Ramsey, Millett & Hudson, 1874.

Morgan, Ted. *Shovel of Stars: The Making of the American West 1800 to the Present.* New York: Touchstone, 1996.

Moser, JoAnn. *Mason Jar Nation: The Jars That Changed America and 50 Clever Ways to Use Them Today.* Minneapolis, MN: Cool Springs Press, 2016.

Nevius, James. "New York's Built Environment Was Shaped by Pandemics." *Curbed*, March 19, 2020. https://ny.curbed.com/2020/3/19/21186665/coronavirus-new -york-public-housing-outbreak-history.

O'Connell, William E. "The Development of the Private Railroad Freight Car, 1830– 1966." *Business History Review* 44, no. 2 (1970): 190–209.

Ogle, Maureen. *In Meat We Trust: An Unexpected History of Carnivore America.* New York: Houghton Mifflin Harcourt, 2013.

Pacyga, Dominic A. *Chicago: A Biography.* Chicago: University of Chicago Press, 2009.

Pierce, Bessie Louise. *History of Chicago, Volume III: The Rise of a Modern City, 1871–1893*. Chicago: University of Chicago Press, 2007.

Roberts, Jeffrey P. *Salted and Cured: Savoring the Culture, Heritage, and Flavor of America's Preserved Meats*. White River Junction, VT: Chelsea Green Publishing, 2017.

Robertson, James I. *After the Civil War: The Heroes, Villains, Soldiers, and Civilians Who Changed America*. Washington, D.C.: National Geographic, 2015.

Rosenberg, Chaim M. *America at the Fair: Chicago's 1893 World's Columbian Exposition*. Charleston, SC: Arcadia, 2008.

Sandburg, Carl. *Chicago Poems: Unabridged*. Mineola, NY: Dover Publications, 2012.

Sandweiss, Martha, Ed Milner, Clyde Milner, and Carol A. O'Connor. *The Oxford History of the American West*. New York: Oxford University Press, 1994.

Smith, Andrew F. *Eating History: 30 Turning Points in the Making of American Cuisine*. New York: Columbia University Press, 2009.

Turner, Katherine Leonard. *How the Other Half Ate: A History of Working-Class Meals at the Turn of the Century*. Berkeley: University of California Press, 2014.

Walsh, Margaret. *The Rise of the Midwestern Meat Packing Industry*. Lexington: University Press of Kentucky, 2021.

Warren, Wilson J. *Tied to the Great Packing Machine: The Midwest and Meatpacking*. Iowa City: University of Iowa Press, 2007.

Weisberger, Bernard A. *The Titans*. Boston: New Word City, 2016.

Wilson, Jeff. *Produce Traffic and Trains*. Waukesha, WI: Kalmbach Books, 2018.

Winans, Charles. "The Evolution of a Vast Industry." *Harper's Weekly*, July 1, 1905.

Youngblood, Dawn. *The SMS Ranch*. Charleston, SC: Arcadia Publishing, Inc., 2017.

Ziegelman, Jane, and Andrew Coe. *A Square Meal: A Culinary History of the Great Depression*. New York: Harper, 2016.

CHAPTER 5

Bonah, Christian, David Cantor, and Mathias Dörries (eds.). *Meat, Medicine and Human Health in the Twentieth Century*. New York: Routledge, 2015.

Boyd, William. "Making Meat: Science, Technology, and American Poultry Production." *Technology and Culture* 42, no. 4 (Oct. 2001): 631–64.

Broadway, Michael J., and Donald D. Stull. *Slaughterhouse Blues: The Meat and Poultry Industry in North America*. Boston: Cengage Learning, 2012.

Bugos, Glenn E. "Intellectual Property Protection in the American Chicken-Breeding Industry." *Business History Review* 66, no. 1 (Spring 1992): 127–68.

Buzzell, Robert D. "Is Vertical Integration Profitable?" *Harvard Business Review*, Jan. 1983. https://hbr.org/1983/01/is-vertical-integration-profitable.

Clark, Thomas D. "The Furnishing and Supply System in Southern Agriculture since 1865." *Journal of Southern History* 12, no. 1 (1946): 24–44.

Cole, Larry. *Communication in Poultry Grower Relations: A Blueprint to Success*. Ames: Iowa State University Press, 2008.

Collingham, Lizzie. *The Taste of War: World War II and the Battle for Food*. New York: Penguin, 2013.

Fite, Gilbert C. *Cotton Fields No More: Southern Agriculture, 1865–1980*. Lexington: University Press of Kentucky, 1984.

Gisolfi, Monica R. *The Takeover: Chicken Farming and the Roots of American Agribusiness*. Athens: University of Georgia Press, 2017.

Golay, Michael. *America 1933: The Great Depression, Lorena Hickok, Eleanor Roosevelt, and the Shaping of the New Deal*. New York: Simon & Schuster, 2016.

Gordon, John Steele. "The Chicken Story." *American Heritage*, Sept. 1996. https://www.americanheritage.com/chicken-story.

Hart, John Fraser. *The Changing Scale of American Agriculture*. Charlottesville: University of Virginia Press, 2003.

Horowitz, Roger. *Food Chains: From Farmyard to Shopping Cart*. Philadelphia: University of Pennsylvania Press, 2011.

Jukes, Thomas, and E. L. Robert Stokstad. "The Multiple Nature of the Animal Protein Factor." *Journal of Biological Chemistry* 180, no. 2 (Sep. 1949): 647–54.

Kahn, Laura H. *One Health and the Politics of Antimicrobial Resistance*. Baltimore: Johns Hopkins University Press, 2016.

Lawler, Andrew. "Chicken of Tomorrow." *Aeon*, Nov. 5, 2014. https://aeon.co/essays/how-the-backyard-bird-became-a-wonder-of-science-and-commerce.

———. "How the Chicken Became Our National Bird." *Saveur*, Sep. 8, 2016. https://www.saveur.com/chicken-of-tomorrow/.

Lawler, Andrew, and Jerry Adler. "How the Chicken Conquered the World." *Smithsonian*, June 2012. https://www.smithsonianmag.com/history/how-the-chicken-conquered-the-world-87583657/.

Leeson, Steven, and John D. Summers. *Broiler Breeder Production*. Nottingham, UK: Nottingham University Press, 2010.

Litwack, Gerald. *Vitamins and Hormones: Advances in Research and Applications, Volume 52*. San Diego: Academic Press, 2011.

Lymbery, Philip. *Farmageddon: The True Cost of Cheap Meat*. London: Bloomsbury, 2015.

Maslowski, Susan. "Farmer's Table: Chicken Lombardy Casserole." *Charleston Gazette-Mail*, June 1, 2016. https://www.wvgazettemail.com/metrokanawha/farmer-s-table-chicken-lombardy-casserole/article_c69d96f4-aac5-5b92-88c6-e35e9479f846.html.

McDowell, Lee R. *Vitamin History, the Early Years*. Sarasota, FL: First Edition Design Publishing, 2013.

McKenna, Maryn. *Big Chicken: The Incredible Story of How Antibiotics Created Modern Agriculture and Changed the Way the World Eats*. Washington, D.C.: National Geographic, 2017.

McKittrick, Meredith. "Industrial Agriculture." In *A Companion to Global Environmental History*, edited by J. R. McNeill and Erin S. Mauldin. West Sussex, UK: Wiley, 2015.

Moody, John. *The Frugal Homesteader: Living the Good Life on Less*. Gabriola Island, BC: New Society Publishers, 2018.

Norval, Mary. "A Short Circular History of Vitamin D from Its Discovery to Its Effects." *Journal of the Royal Medical Society* 268, no. 2 (2005).

Ogle, Maureen. *In Meat We Trust: An Unexpected History of Carnivore America.* New York: Houghton Mifflin Harcourt, 2013.

Pappworth, Maurice Henry. "Vitamin B12 (from *Streptomyces Griseus*) in Pernicious Anaemia." *British Medical Journal* 1, no. 4665 (June 1950): 1302–3.

Poultry Science Association. "Annual Meeting." *Poultry Science* 31, no. 1 (Jan. 1952): 10.

Ravina, Enrique. *The Evolution of Drug Discovery: From Traditional Medicines to Modern Drugs.* Weinheim, Germany: Wiley, 2011.

Roberts, Paul. *The End of Food.* New York: Mariner Books, 2009.

Robison, W. L. *Vitamin B-12 Supplements for Growing and Fattening Pigs.* Ohio Agricultural Experiment Station. April, 1953. https://kb.osu .edu/bitstream/handle/1811/63024/OARDC_research_bulletin_n0729. pdf?sequence=1&isAllowed=y.

Rosenfeld, Louis. "Vitamine—Vitamin. The Early Years of Discovery." *Clinical Chemistry* 43, no. 4 (April 1997): 680–85.

Rude, Emelyn. *Tastes Like Chicken: A History of America's Favorite Bird.* New York: Pegasus Books, 2016.

Sachs, Jessica Snyder. *Good Germs, Bad Germs: Health and Survival in a Bacterial World.* New York: Hill and Wang, 2007.

Sawyer, Gordon. *Northeast Georgia: A History.* Charleston, SC: Arcadia, 2001.

Schrepfer, Susan R., and Philip Scranton (eds.). *Industrializing Organisms: Introducing Evolutionary History.* New York: Routledge, 2004.

Silbergeld, Ellen K. *Chickenizing Farms and Food: How Industrial Meat Production Endangers Workers, Animals, and Consumers.* Baltimore: Johns Hopkins University Press, 2016.

Smith, Andrew F. *Food in America: The Past, Present, and Future of Food, Farming, and the Family Meal.* Santa Barbara, CA: ABC-CLIO, 2017.

Stuesse, Angela. *Scratching Out a Living: Latinos, Race, and Work in the Deep South.* Oakland: University of California Press, 2016.

Thompson, Gabriel. *Working in the Shadows: A Year of Doing the Jobs (Most) Americans Won't Do.* New York: Nation Books, 2011.

Twilley, Nicola, and Cynthia Graber. "How the Chicken Industry Got Hooked on Antibiotics." *The Atlantic*, Aug. 16, 2017. https://www.theatlantic.com/science/archive/2017/08/big-pharma-big-chicken/536979/.

Urry, Amelia. "Our Crazy Farm Subsidies, Explained." *Grist*, April 20, 2015. https://grist.org/food/our-crazy-farm-subsidies-explained/.

Vardeman, Johnny. "Johnny Vardeman: Storm Victim's Cemetery Marker in Need of Repair." *Gainseville Times*, March 25, 2017. https://www.gainesvilletimes.com/columnists/columnists/johnny-vardeman-storm-victims-cemetery-marker-in-need-of-repair/.

Weinberg, Carl. "Jesse Jewell (1902–1975)." *New Georgia Encyclopedia*. Jan. 14, 2005. https://www.georgiaencyclopedia.org/articles/business-economy/jesse-jewell-1902-1975.

Williams, Tara Layman. *The Complete Guide to Raising Chickens: Everything You Need to Know Explained Simply*. Ocala, FL: Atlantic Publishing Group, 2011.

Williams, William Henry. *Delmarva's Chicken Industry: 75 Years of Progress*. Georgetown, DE: Delmarva Poultry Industry, Inc., 1998.

Wright, Gavin, and Howard Kunreuther. "Cotton, Corn and Risk in the Nineteenth Century." *Journal of Economic History* 35, no. 3 (1975): 526–51.

CHAPTER 6

Anderson, Avis H. *A & P: The Story of the Great Atlantic & Pacific Tea Company*. Charleston, SC: Arcadia, 2002.

Bartholomew, Ivy A. "Matters of Taste: Spam Mania." *The Rotarian*, August 2000.

Batchelor, Bob. *American Pop: Popular Culture Decade by Decade*. Westport, CT: Greenwood, 2008.

Belluz, Julia. "The Average American Woman Now Weighs 166 Pounds— As Much as a 1960s Man." *Vox*. Updated June 17, 2015. https://www.vox.com/2015/6/15/8784389/america-weight-gain.

Black, Dustin, Armstrong, Dan. *The Book of Spam: A Most Glorious and Definitive Compendium of the World's Favorite Canned Meat*. New York: Atria Books, 2008.

Bromell, Nicolas. "The Automat: Preparing the Way for Fast Food." *New York History* 81, no. 3 (2000): 300–312.

BuzzFeed Multiplayer. "McDonald's: 1955 vs. Now." YouTube video, 4:40, Sep. 4, 2015. https://www.youtube.com/watch?v=oqI4rR6lXU0.

Carraher, Charles E. Jr. *Seymour/Carraher's Polymer Chemistry: Sixth Edition.* Boca Raton, FL: CRC Press, 2003.

Chibuzo. "How Smart Thinking Drove the Initial Growth of A&P." High Impact Nation, Oct. 1, 2021. https://highimpactnation.net/2021/01/10/how-smart-thinking-drove-the-initial-growth-of-ap/.

Critser, Greg. *Fat Land: How Americans Became the Fattest People in the World.* New York: Houghton Mifflin Harcourt, 2004.

Deck, Cecilia. "Fast-Food Pioneer A&W Survives to Map Comeback." *Chicago Tribune,* Nov. 19, 1989. https://www.chicagotribune.com/news/ct-xpm-1989-11-19-8903110146-story.html.

Edwards, Owen. "How 260 Tons of Thanksgiving Leftovers Gave Birth to an Industry." *Smithsonian,* Dec. 2004. https://www.smithsonianmag.com/history/tray-bon-96872641/.

Ellickson, Paul B. "The Evolution of the Supermarket Industry: From A&P to Wal-Mart." In *Handbook on the Economics of Retail and Distribution,* edited by Emek Basker. Cheltenham, UK: Edward Elgar Publishing, 2016.

Faulk, Richard. *The Next Big Thing: A History of the Boom-or-Bust Moments That Shaped the Modern World.* San Francisco: Zest Books, 2019.

Ferdman, Roberto. "America's 60-Year Love Affair with Frozen TV Dinners Is Over." *Quartz,* March 13, 2014. https://qz.com/187433/americas-60-year-love-affair-with-frozen-tv-dinners-is-over/.

Gantz, Carroll. *Refrigeration: A History.* Jefferson, NC: McFarland, 2015.

Genoways, Ted. *The Chain: Farm, Factory, and the Fate of Our Food.* New York: HarperCollins, 2015.

Gref, Lynn G. *The Rise and Fall of American Technology.* New York: Algora, 2010.

Grimes, William. "Michael James Delligatti, Creator of the Big Mac, Dies at 98." *New York Times,* Nov. 30, 2016. https://www.nytimes.com/2016/11/30/business/michael-james-delligatti-creator-of-the-big-mac-dies-at-98.html.

Hamilton, Andrew. "Dinners without Drudgery." *Popular Mechanics* 87, no. 4 (April 1947): 174–77.

Hamilton, Shane. "The Economies and Conveniences of Modern-Day Living: Frozen Foods and Mass Marketing, 1945–1965." *Business History Review* 77, no. 1 (2003): 33–60.

Harris, William. "10 Most Popular McDonald's Menu Items of All Time." HowStuffWorks, April 7, 2009. https://money.howstuffworks.com/10-popular -mcdonalds-menu-items.htm#pt4.

Hisano, Ai. "Cellophane, the New Visuality, and the Creation of Self-Service Food Retailing." Harvard Business School Working Paper, No. 17-106, May 2017.

Hogan, David G. *Selling 'em by the Sack: White Castle and the Creation of American Food.* New York: New York University Press, 1999.

Jackson, Tom. *Chilled: How Refrigeration Changed the World and Might Do So Again.* London: Bloomsbury, 2015.

Jakle, John A., and Keith A. Sculle. *Fast Food: Roadside Restaurants in the Automobile Age.* Baltimore: Johns Hopkins University Press, 2002.

Kieler, Ashlee. "The White Castle Story: The Birth of Fast Food and the Burger Revolution." *Consumerist*, July 14, 2015. https://consumerist.com/2015/07/14/the -white-castle-story-the-birth-of-fast-food-the-burger-revolution/.

Kurlansky, Mark. *Frozen in Time: Clarence Birdseye's Outrageous Idea about Frozen Food.* New York: Delacorte Press, 2014.

Leonard, Christopher. *The Meat Racket: The Secret Takeover of America's Food Business.* New York: Simon & Schuster, 2015.

Levinson, Marc. *The Great A&P and the Struggle for Small Business in America.* New York: Hill and Wang, 2011.

Love, John F. *McDonald's: Behind the Arches.* New York: Bantam Books, 1995.

New York Times. "TV Dinners Seek Gourmet Market." Feb. 10, 1984. https://www .nytimes.com/1984/02/10/business/tv-dinners-seek-gourmet-market.html.

Nibert, David Alan. *Animal Oppression and Human Violence: Domesecration, Capitalism, and Global Conflict.* New York: Columbia University Press, 2013.

Novak, Matt. "The Great Depression and the Rise of the Refrigerator." *Pacific Standard.* Updated June 14, 2017. https://psmag.com/environment/the-rise-of-the -refrigerator-47924.

Rees, Jonathan. *Refrigeration Nation: A History of Ice, Appliances, and Enterprise in America*. Baltimore: Johns Hopkins University Press, 2013.

Risch, Erna. *The Quartermaster Corps: Organization, Supply, and Services Volume I*. Washington, D.C.: Center of Military History, United States Army, 1995.

Rosenwald, Michael S. "From the A&P to Amazon: The Rise of the Modern Grocery Store." *Washington Post*, June 16, 2017. https://www.washingtonpost.com/news/retropolis/wp/2017/06/16/from-the-ap-to-amazon-the-rise-of-the-modern-grocery-store/.

Rude, Emelyn. *Tastes Like Chicken: A History of America's Favorite Bird*. New York: Pegasus Books, 2016.

Ruhlman, Michael. *Grocery: The Buying and Selling of Food in America*. New York: Abrams, 2018.

Schlosser, Eric. *Fast Food Nation: The Dark Side of the All-American Meal*. New York: Mariner Books, 2012.

Smith, Andrew F. (ed.). *The Oxford Encyclopedia of Food and Drink in America*. New York: Oxford University Press, 2013.

Smith, Andrew F. *Savoring Gotham: A Food Lover's Companion to New York City*. New York: Oxford University Press, 2015.

Smith, Andrew F. *The Turkey: An American Story*. Champaign: University of Illinois Press, 2006.

Smith, Annabelle K. "The Strange History of Frozen Food." *Eater*, Aug. 21, 2014. https://www.eater.com/2014/8/21/6214423/the-strange-history-of-frozen-food-from-clarence-birdseye-to-the.

Tweedie, Steven. "How the Microwave Was Invented by a Radar Engineer Who Accidentally Cooked a Candy Bar in His Pocket." *Business Insider*, July 3, 2015. https://www.businessinsider.com/how-the-microwave-oven-was-invented-by-accident-2015-4.

CHAPTER 7

American Heart Association Editorial Staff. "Saturated Fat." https://www.heart.org/en/healthy-living/healthy-eating/eat-smart/fats/saturated-fats.

Duke University Medical Center. "Salt Appetite Is Linked to Drug Addiction, Research Finds." ScienceDaily. http://www.sciencedaily.com/releases/2011/07/110711151451.htm.

Endocrine Today. "With 'Bliss Points' and 'Mouth Feel,' Food Industry Plays Role in Hedonic Eating Habits." *Healio*, Nov. 14, 2018. https://www.healio.com/endocrinology/obesity/news/online/%7B0c3a46c9-3ebc-4fee-a5aa-c8ddb4e37dac%7D/with-bliss-points-and-mouth-feel-food-industry-plays-role-in-hedonic-eating-habits.

McQuaid, John. *Tasty: The Art and Science of What We Eat.* New York: Scribner, 2015.

Moorhead, Alana. "Slice of Heaven: Ever Wondered Why Toast Tastes So Much Better Than Bread? There's Actually a Scientific Reason Behind It." *The Sun*, June 26, 2016. https://www.thesun.co.uk/living/1344052/ever-wondered-why-toast-tastes-so-much-better-than-bread-theres-actually-a-scientific-reason-behind-it/.

Mouritsen, Ole G., and Klavs Styrbæk. *Umami: Unlocking the Secrets of the Fifth Taste.* New York: Columbia University Press, 2014.

O'Connell, Libby. *The American Plate: A Culinary History in 100 Bites.* Naperville, IL: Sourcebooks, 2014.

Squire, Larry R. (ed.). *Encyclopedia of Neuroscience.* Cambridge, MA: Elsevier/Academic Press, 2009.

U.S. Food and Drug Administration. "Added Sugars on the New Nutrition Facts Label." FDA.gov. Updated March 11, 2020. https://www.fda.gov/food/new-nutrition-facts-label/added-sugars-new-nutrition-facts-label.

Zaraska, Marta. *Meathooked: The History and Science of Our 2.5-Million-Year Obsession with Meat.* New York: Basic Books, 2016.

CHAPTER 8

Batheja, Aman. "The Time Oprah Winfrey Beefed with the Texas Cattle Industry." *Texas Tribune*, Jan. 10, 2018. https://www.texastribune.org/2018/01/10/time-oprah-winfrey-beefed-texas-cattle-industry/.

Cook, Ken. "Government's Continuing Bailout of Corporate Agriculture." Environmental Working Group, May 5, 2010. https://www.ewg.org/news -insights/news/governments-continuing-bailout-corporate-agriculture.

Coudray, Guillaume. *Who Poisoned Your Bacon? The Dangerous History of Meat Additives*. London: Icon Books, 2021.

Davies, Steve. "Court Dismisses Challenge to Payments for Pork Trademark." *Agri-Pulse*, Aug. 23, 2019. https://www.agri-pulse.com/articles/12538-court-dismisses -challenge-to-payments-for-pork-trademark.

Genoways, Ted. "Gagged by Big Ag." *Mother Jones*, July 2013. https://www.mother jones.com/environment/2013/06/ag-gag-laws-mowmar-farms/.

Lusk, Jayson. "Are Farm Subsidies Making Us Fat?" *Jayson Lusk: Food and Agricultural Economist* (blog), Aug. 11, 2021. http://jaysonlusk.com/ blog/2016/7/22/are-farm-subsidies-making-us-fat.

Morgan, Dan. "Industry Finds a Way around Budget Cutters." *Washington Post*, June 26, 1995. https://www.washingtonpost.com/archive/politics/1995/06/26/ industry-finds-a-way-around-budget-cutters/ff605843-4532-4e95-ad08- 88bb590b3e6f/.

Nepveux, Michael. "Chicken and Pork in Cold Storage Lead to Lower Overall Levels of Meat and Poultry." American Farm Bureau, March 24, 2021. https://www.fb .org/market-intel/chicken-and-pork-in-cold-storage-lead-to-lower-overall-levels -of-meat-and-p.

Nestle, Marion. "Least Credible Food Industry Ad of the Week: JBS and Climate Change." *Food Politics* (blog), April 26, 2021. https://www.foodpolitics .com/2021/04/least-credible-food-industry-ad-of-the-week-jbs-and-climate -change/.

US Agricultural Marketing Service. "USDA Additional Food Purchase Plans." US Department of Agriculture, May 4, 2020. https://www.ams.usda.gov/press-release/ usda-announces-additional-food-purchase-plans.

US Department of Agriculture. "Regulatory Reform at a Glance. Proposed Rule: School Meals Flexibilities." Jan. 2020. https://fns-prod.azureedge.net/sites/default/ files/resource-files/school-meals-flexibilities-fact-sheet.pdf.

Vinik, Danny. "A $60 Million Pork Kickback?" *Politico*, Aug. 30, 2015. https://www .politico.com/agenda/story/2015/08/a-60-million-pork-kickback-000210/.

Wildavsky, Aaron. *But Is It True? A Citizen's Guide to Environmental Health and Safety Issues.* Cambridge, MA: Harvard University Press, 1997.

CHAPTER 9

Brissette, Christy. "Is Meat Manly? How Society Pressures Us to Make Gendered Food Choices." *Washington Post*, Jan. 25, 2017. https://www.washingtonpost.com/lifestyle/wellness/is-meat-manly-how-society-pressures-us-to-make-gendered-food-choices/2017/01/24/84669506-dce1-11e6-918c-99ede3c8cafa_story.html.

Casserly, Meghan. "Grilling, Guys and the Great Gender Divide." *Forbes*, July 1, 2010. https://www.forbes.com/2010/07/01/grilling-men-women-barbecue-forbes-woman-time-cooking.html?sh=6207f7ecbad6.

Facts First (blog). "Does Biden's Climate Plan Include 'Cutting 90% of Red Meat from Our Diets by 2030?'" CNN. https://www.cnn.com/factsfirst/politics/factcheck_e5e088b0-0b69-400b-aa5d-b5cfb9168d33.

Goodyer, Paula. "Meat Eaters Justify Diet Using 'Four Ns': Natural, Necessary, Normal, Nice." *Sydney Morning Herald*, May 30, 2015. https://www.smh.com.au/lifestyle/health-and-wellness/meat-eaters-justify-diet-using-four-ns-natural-necessary-normal-nice-20150530-ghd5le.html.

Hendricks, Scotty. "Poorer People Eat More Meat to Feel Affluent, New Study Claims." *Big Think*, Sep. 13, 2018. https://bigthink.com/scotty-hendricks/a-new-study-finds-that-poorer-people-eat-more-meat-to-feel-more-affluent.

Hodson, Gordon. "The Meat Paradox: Loving but Exploiting Animals." *Psychology Today*, March 3, 2014. https://www.psychologytoday.com/us/blog/without-prejudice/201403/the-meat-paradox-loving-exploiting-animals.

Ramanujan, Krishna. "Eating Green Could Be in Your Genes." *Cornell Chronicle*, March 29, 2016. https://news.cornell.edu/stories/2016/03/eating-green-could-be-your-genes.

Smith, Kat. "Conservatives Consume Way More Meat and Dairy Than Liberals, Study Finds." LiveKindly. https://www.livekindly.co/conservatives-more-meat-than-liberals/.

Sparkman, Gregg, et al. "Cut Back or Give It Up? The Effectiveness of Reduce and Eliminate Appeals and Dynamic Norm Messaging to Curb Meat Consumption." *Journal of Environmental Psychology* 75 (June 2021): 101592.

Sugar, Rachel. "The Politics of 'Dude Food.'" *Vox*, Jan. 6, 2021. https://www.vox
.com/the-goods/22178806/diners-dudes-diets-emily-contois.

University of Oslo. "The Meat Paradox." ScienceDaily, Oct. 11, 2016. www
.sciencedaily.com/releases/2016/10/161011125655.htm.

Weill, Kelly. "Why Right Wingers Are Going Crazy about Meat." *Daily Beast*, Aug.
25, 2018. https://www.thedailybeast.com/why-right-wingers-are-going-crazy
-about-meat.

Whalen, Eamon. "How Red Meat Became the Red Pill for the Alt-Right." *The
Nation*, June 15, 2020. https://www.thenation.com/article/society/beef-red-pill
-right/.

Zaraska, Marta. "Hooked on Meat: How Cultural Beliefs and Attitudes Drive
Meat Consumption." *Meatonomics* (blog), April 20, 2016. https://meatonomics
.com/2016/04/20/hooked-on-meat-how-cultural-beliefs-and-attitudes-drive-meat
-consumption/.

CHAPTER 10

Agricultural Marketing Service. "National Monthly Grass Fed Beef Report." US
Department of Agriculture, Aug. 27, 2021. https://www.ams.usda.gov/mnreports/
lsmngfbeef.pdf.

Benenson, Bob. "Grazing in the Grass Is Growing Fast." New Hope Network,
April 3, 2019. https://www.newhope.com/market-data-and-analysis/grazing-grass
-growing-fast.

Brissette, Christy. "Your 'Grass-Fed' Beef May Not Have Come from a Cow
Grazing in a Pasture. Here's Why." *Washington Post*, Dec. 19, 2018. https://www
.washingtonpost.com/lifestyle/wellness/your-grass-fed-beef-may-have-not-have
-come-from-a-cow-grazing-in-a-pasture-heres-why/2018/12/13/7e65ebb2-fc91
-11e8-83c0-b06139e540e5_story.html.

Consumer Reports. "Why Grass-Fed Beef Costs More." *Yahoo Finance*, Aug. 24,
2015. https://finance.yahoo.com/news/why-grass-fed-beef-costs-100000222.html.

Goldman, Jason G. "Is Meat-Eating a Conversation Tactic?" *The Thoughtful Animal*
(blog), April 12, 2013, https://blogs.scientificamerican.com/thoughtful-animal/is
-meat-eating-a-conservation-tactic/.

Held, Lisa. "Despite Many Challenges, Grassfed Beef Could Go Mainstream." *Civil Eats*, June 7, 2017. https://civileats.com/2017/06/07/despite-many-challenges -grassfed-beef-could-go-mainstream/.

Humaneitarian. "On Being a Humaneitarian." https://humaneitarian.org/being -a-humaneitarian/on-being-a-humaneitarian/#.YSrlBS1h3BB.

LowImpact.org. "Is Eating Meat Ethical or Sustainable? Interview with Simon Fairlie, Author of '*Meat: A Benign Extravagance.*'" Sep. 16, 2018. https://www .lowimpact.org/is-eating-meat-ethical-simon-fairlie-interview/.

New York Times. "Your Questions about Food and Climate Change, Answered." April 30, 2019. https://www.nytimes.com/interactive/2019/04/30/dining/climate -change-food-eating-habits.html?smid=fb-share.

O'Connor, Jennifer. "Barriers for Farmers and Ranchers to Adopt Regenerative Ag Practices in the U.S." Guidelight Strategies, Aug. 2020. https://forainitiative.org/ wp-content/uploads/Barriers-to-Adopt-Regnerative-Agriculture-Interactive.pdf.

Skool of Vegan. "I Only Eat Humane Animal Products." http://www.skoolofvegan .com/humane-meat.html.

Stone Barns Center for Food and Agriculture. "Back to Grass: The Market Potential for U.S. Grassfed Beef." April, 2017. https://www.organicconsumers.org/sites/ default/files/grassfed-marketstudy-f.pdf.

White Oak Pastures. "Study: White Oak Pastures Beef Reduces Atmospheric Carbon." June 4, 2019. https://blog.whiteoakpastures.com/blog/carbon-negative -grassfed-beef.

Zimberoff, Larissa. "There's a New 'Organic' Food That Fights Global Warming." *Bloomberg Green*, April 23, 2021. https://www.bloomberg.com/news/ articles/2021-04-23/regenerative-farming-is-a-new-kind-of-organic-food-that -s-good-for-earth-too.

CHAPTER 11

Balu, Nivedita. "Impossible Foods Raises $200 Million in Fresh Funding." Reuters, Aug. 13, 2020. https://www.reuters.com/article/us-impossible-foods-funding/ impossible-foods-raises-200-million-in-fresh-funding-idUSKCN2592WV.

Plant Based Foods Association. "2020 Retail Sales Data Announcement." Press release, April 6, 2021. https://www.plantbasedfoods.org/2020-retail-sales-data -announcement/.

Shanker, Deena. "Impossible and Beyond Slash Prices as Fake-Meat Market Heats Up." *Bloomberg Businessweek*, April 16, 2021. https://www.bloomberg.com/news/ articles/2021-04-16/beyond-meat-bynd-impossible-foods-battle-over-future-of -fake-meat-industry.

CHAPTER 12

Blaustein-Rejto, Dan, and Alex Smith. "We're on Track to Set a New Record for Global Meat Consumption." *MIT Technology Review*, April 26, 2021. https:// www.technologyreview.com/2021/04/26/1023636/sustainable-meat-livestock -production-climate-change/.

Flynn, Adam. "Industry Parallels: Algal Biofuels." New Harvest 2018 Conference, July 21, 2018. https://www.pscp.tv/futurefoodshow/1ZkKzNvMNqwKv.

University of Oxford. "Lab-Grown Meat Would Cut Greenhouse Gas Emissions and Save Energy, Research Suggests." ScienceDaily, July 18, 2011. http://www .sciencedaily.com/releases/2011/07/110714101036.htm.

Zaraska, Marta. "Lab-Grown Meat Is in Your Future, and It May Be Healthier Than the Real Stuff." *Washington Post*, May 2, 2016. https://www.washingtonpost .com/national/health-science/lab-grown-meat-is-in-your-future-and-it-may-be -healthier-than-the-real-stuff/2016/05/02/aa893f34-e630-11e5-a6f3-21ccdbc5 f74e_story.html.

EPILOGUE

Lusk, Jayson. "The Political Polarization of Meat Demand." *Jayson Lusk: Food and Agricultural Economist* (blog), April 23, 2019. http://jaysonlusk.com/ blog/2019/4/23/the-political-polarization-of-meat-demand.

Acknowledgments

Meat Me Halfway—and the reducetarian movement from which it was born—is the culmination of a near decade of benefiting from the insights, kindness, and hard work of literally thousands of people I have come to deeply admire. This book is for and because of you.

I am enormously grateful for my agent, Linda Konner, who has proved to be a fierce alley in my literary pursuits. Few writers are as lucky as me to have someone who is as skilled and honest as you in their corners.

To Jake Bonar: Your enthusiasm for the book on behalf of Globe Pequot–Prometheus Books means the world to me. Thank you for believing in the promise of *Meat Me Halfway* the moment it came across your desk. And kudos to Jessie McCleary and Jess Kastner for their assistance with producing, marketing, and publicizing the book.

I owe a great deal to Nicholas Bromley and Rhys Southan, whose editing skills are unparalleled, and the book is surely more concise and reasonable because of it.

It's hard for me to overstate the enormous influence Sofia Davis-Fogel has had on the way I think, and the way I think about the world.

Thank you to Emily Byrd for making the manuscript as factual as possible.

I'm grateful for Chris Davidson and the eye-catching book jacket he designed. You are a supremely talented graphic designer.

To Journey Wade-Hak, who put his blood, sweat, and tears into the documentary version of *Meat Me Halfway*, much of which is a part of this book: You are not only a talented director, but also an enormously decent human. You are the Rick to my Morty.

Thank you to all those who contributed quotes to this book. Each of you are a wealth of knowledge and I so appreciate your expertise and willingness to share your wisdom with me.

I'm especially grateful for my mom and dad, who add color to my life and every single one of these pages. I am in your debt for loving me unconditionally.

To Tobey and Cooper: Thank you for all the snuggles and licks. My life is more silly, joyful, and calm because you two nugglets are in it.

To all of my dearest friends (including but not limited to Dan Feldman, Mike Young, David DiLillo, Danielle Medina, Vincent Romano, Arthur Kapetanakis, Joe Eastman, and Michael Trollan): Thank you for making me laugh whenever I needed it most.

And to Isabel, my wife, my best friend, my everything: I love you always and forever. It's as simple as that.